Mathematics of Engineering and Science

Mehdi Rahmani-Andebili

Mathematics of Engineering and Science

Practice Problems, Methods, and Solutions

 Springer

Mehdi Rahmani-Andebili
ECE Department
University of Alabama
Tuscaloosa, AL, USA

ISBN 978-3-031-71933-2   ISBN 978-3-031-71934-9   (eBook)
https://doi.org/10.1007/978-3-031-71934-9

This Springer imprint is published by the registered company Springer Nature Switzerland AG
The registered company address is: Gewerbestrasse 11, 6330 Cham, Switzerland

If disposing of this product, please recycle the paper.

Preface

The Engineering Mathematics and Mathematical Methods in Sciences are the necessary courses for all engineering and science majors, respectively, that are taught at universities and colleges worldwide. This textbook has been prepared for instructors as well as for students taking these courses. In each chapter of the textbook, different types of problems and exercises have been presented that are categorized as follows.

- *Problems with detailed solution*: They have been designed to teach students the subjects in detail. Moreover, they have been categorized in different levels based on their difficulty levels (easy, normal, and hard) and calculation amounts (small, normal, and large). These classifications can help students study the book in the most efficient way.
- *Partially solved exercises*: They have been designed to encourage students to practice more problems while guiding them through the problem-solving procedure and hinting the required formulas.
- *Exercises with final answer*: They have been designed to encourage students to practice by themselves while hinting them by the final answer as well as to help instructors to give tests or quizzes.

In the following, the description of each chapter is briefly presented.

Chapters 1 and 2 cover the subjects concerned with the complex quantities, limit of complex functions, complex equations, and holomorphic functions and their harmonic conjugate functions.

Chapters 3 and 4 teach the complex transformations that include linear, power, reciprocal, exponential, natural logarithm, hyperbolic sine and cosine, sine and cosine, and linear fractional complex transformations.

Chapters 5 and 6 study the singularities of complex functions (such as poles, removable singularity, and essential singularities), complex series, Taylor and Laurent series expansions of complex functions, and residue of complex functions.

Chapters 7 and 8 review different complex integrations that include complex integration of nonholomorphic functions, complex integration of holomorphic functions, and complex integration of functions including finite number of singular points.

Chapters 9 and 10 investigate the Fourier series of periodic functions, half-domain Fourier sine and cosine series of aperiodic functions, complex Fourier series of periodic functions, Fourier integral of aperiodic functions, complex Fourier integral of aperiodic functions, Fourier transform of aperiodic functions, and half-domain Fourier sine and cosine transforms of aperiodic functions.

Chapters 11 and 12 are concerned with the determination of type of partial differential equations, updating partial differential equations by new variables and solving the partial differential equations. In this regard, the partial differential equations are solved by using several techniques such as the approaches used in solving ordinary differential equations, the method of characteristics equation, the technique of variables separation, and Laplace transform. Moreover, solving partial differential equations in the steady-state condition is presented.

If the students have difficulties in studying the textbook, they can first study the calculus series books covering Precalculus, Calculus I, Calculus II, and Calculus III. The subjects of the calculus series books are as follows.

Calculus III: Practice Problems, Methods, and Solutions, Springer Nature, 2023.
- *Linear Algebra and Analytical Geometry*
- *Lines, Surfaces, and Vector Functions in Three-Dimensional Coordinate System*
- *Multivariable Functions*
- *Double Integrals and Their Applications*
- *Triple Integrals and Their Applications*
- *Line Integrals and Their Applications*

Calculus II: Practice Problems, Methods, and Solutions, Springer Nature, 2023.
- *Applications of Integration*
- *Sequences and Series and Their Applications*
- *Polar Coordinate System*
- *Complex Numbers*

Calculus I: Practice Problems, Methods, and Solutions, Springer Nature, 2023.
- *Characteristics of Functions*
- *Trigonometric Equations and Identities*
- *Limits and Continuities*
- *Derivatives and Their Applications*
- *Definite and Indefinite Integrals*

Precalculus: Practice Problems, Methods, and Solutions, Springer Nature, 2024.
- *Real Number Systems, Exponents and Radicals, and Absolute Values and Inequalities*
- *Systems of Equations*
- *Quadratic Equations*
- *Functions, Algebra of Functions, and Inverse Functions*
- *Factorization of Polynomials*
- *Trigonometric and Inverse Trigonometric Functions*
- *Arithmetic and Geometric Sequences*

Since the textbook includes the basic and advanced problems with very detailed problem solutions, it can be used as a practicing study guide by students and as a supplementary teaching source by instructors. Moreover, since the problems and exercises have very detailed solutions, the textbook is helpful for under-prepared students. In addition, it is beneficial for knowledge-able students because it includes advanced problems and exercises.

In preparing the problems and solutions, care has been taken to use methods typically found in the primary instructor-recommended textbooks. By considering this key point, the textbook is in the direction of instructors' lectures, and the instructors will not see any untaught and unusual problem solutions in their students' answer sheets.

Tuscaloosa, AL, USA   Mehdi Rahmani-Andebili

The Other Works Published by the Author

The author has already published the books and textbooks below with Springer Nature.

1. MATLAB Lessons, Examples, and Exercises - A Tutorial for Beginners and Experts, *Springer Nature*, 2024.

2. Power System Analysis – Comprehensive Lessons, *Springer Nature*, 2024.

3. Power System Analysis – Practice Problems, Methods, and Solutions, *Springer Nature*, 2022.

4. Feedback Control Systems Analysis and Design – Practice Problems, Methods, and Solutions, *Springer Nature*, 2022.

5. AC Electric Machines – Practice Problems, Methods, and Solutions, *Springer Nature*, 2022.

6. DC Electric Machines, Electromechanical Energy Conversion Principles, and Magnetic Circuit Analysis – Practice Problems, Methods, and Solutions, *Springer Nature*, 2022.

7. Advanced Electrical Circuit Analysis – Practice Problems, Methods, and Solutions, *Springer Nature*, 2022.

8. AC Electrical Circuit Analysis – Practice Problems, Methods, and Solutions, *Springer Nature*, 2021.

9. DC Electrical Circuit Analysis – Practice Problems, Methods, and Solutions, *Springer Nature*, 2020.

10. Differential Equations – Practice Problems, Methods, and Solutions, *Springer Nature*, 2022.

11. Calculus III – Practice Problems, Methods, and Solutions, *Springer Nature*, 2023.

12. Calculus II – Practice Problems, Methods, and Solutions, *Springer Nature*, 2023.

13. Calculus I (2nd Ed.) – Practice Problems, Methods, and Solutions, *Springer Nature*, 2023.

14. Precalculus (2nd Ed.) – Practice Problems, Methods, and Solutions, *Springer Nature*, 2024.

15. Calculus – Practice Problems, Methods, and Solutions, *Springer Nature*, 2021.

16. Precalculus – Practice Problems, Methods, and Solutions, *Springer Nature*, 2021.

17. Planning and Operation of Electric Vehicles in Smart Grid, *Springer Nature*, 2023.

18. Applications of Artificial Intelligence in Planning and Operation of Smart Grid, *Springer Nature*, 2022.

19. Design, Control, and Operation of Microgrids in Smart Grids, *Springer Nature*, 2021.

20. Applications of Fuzzy Logic in Planning and Operation of Smart Grids, *Springer Nature*, 2021.

21. Operation of Smart Homes, *Springer Nature*, 2021.

22. Planning and Operation of Plug-in Electric Vehicles: Technical, Geographical, and Social Aspects, *Springer Nature*, 2019.

Contents

Abstract

In this chapter, the basic and advanced problems of complex quantities, limit of complex functions, complex equations, and holomorphic functions along with their harmonic conjugate functions are studied. Herein, different types of problems and exercises are presented that are categorized as follows:

○ *Problems with detailed solution*: They have been designed to teach students the subjects in detail. Moreover, they have been categorized into different levels based on their difficulty levels (easy, normal, and hard) and calculation amounts (small, normal, and large).

○ *Partially solved exercises*: They have been designed to encourage students to practice more problems while guiding them through the problem-solving procedure and hinting the required formulas as well as to help instructors to give tests or quizzes.

○ *Exercises with final answer*: They have been designed to encourage students to practice by themselves while hinting them by the final answer.

1.1   Complex Quantities

1.1. Calculate the value of $f(z) = (x^2 - y^2) + i(x + y^2)$ for $z = 1 + 2i$ [1–7].

Difficulty level   ● Easy   ○ Normal   ○ Hard
Calculation amount   ● Small   ○ Normal   ○ Large

1) $3 - 5i$
2) $-5 + 3i$
3) $-3 + 5i$
4) $5 - 3i$

Partially Solved Exercise

Calculate the value of $f(i)$ if $f(z) = (x^2 + y^2) + ixy^2$.

Solution

From $z = x + iy = i$, it is noticed that $x = 0$ and $y = 1$. Hence:

© The Author(s), under exclusive license to Springer Nature Switzerland AG 2025
M. Rahmani-Andebili, *Mathematics of Engineering and Science*, https://doi.org/10.1007/978-3-031-71934-9_1

$$f(i) = \left(0^2 + 1^2\right) + i(0)(1)^2$$

$$\Rightarrow f(i) =$$

1.2. Calculate the value of $z^{100}(1 - i)$ for $z = \frac{\sqrt{2}}{2} + i\frac{\sqrt{2}}{2}$.

  Difficulty level　　　○ Easy　● Normal　○ Hard
  Calculation amount　● Small　○ Normal　○ Large

1) $-1 - i$
2) $1 - i$
3) $-1 + i$
4) $1 + i$

1.3. Which one of the following choices is one of the fourth roots of $-9i$?

  Difficulty level　　　○ Easy　● Normal　○ Hard
  Calculation amount　● Small　○ Normal　○ Large

1) $\sqrt{3}\left(\cos\dfrac{5\pi}{8} + i\sin\dfrac{5\pi}{8}\right)$

2) $\sqrt{3}\left(\cos\dfrac{9\pi}{8} + i\sin\dfrac{9\pi}{8}\right)$

3) $\sqrt{3}\left(\cos\dfrac{11\pi}{8} + i\sin\dfrac{11\pi}{8}\right)$

4) $\sqrt{3}\left(\cos\dfrac{13\pi}{8} + i\sin\dfrac{13\pi}{8}\right)$

Partially Solved Exercise

Calculate the sixth roots of $z = 1$.

Solution

The complex quantity can be presented in polar form as follows:

$$1 = 1e^{i0}$$

Now, the sixth roots of the complex quantity can be calculated as follows:

$$z_k = \sqrt[6]{z} = \sqrt[6]{1}\left[\cos\left(\frac{2k\pi + 0}{6}\right) + i\sin\left(\frac{2k\pi + 0}{6}\right)\right], \quad k = 0, 1, \ldots, 5$$

$$z_k = \cos\left(-\!\!-\!\!-\right) + i\sin\left(-\!\!-\!\!-\right), \quad k = 0, 1, \ldots, 5$$

The roots include:

$$z_0 = \cos 0 + i\sin 0 = 1$$

$$z_1 = \cos\frac{\pi}{3} + i\sin\frac{\pi}{3} = \frac{1}{2} + i\frac{\sqrt{3}}{2}$$

$$z_2 = \cos\frac{2\pi}{3} + i\sin\frac{2\pi}{3} = -\frac{1}{2} + i\frac{\sqrt{3}}{2}$$

$$z_3 = \cos\frac{3\pi}{3} + i\sin\frac{3\pi}{3} = -1$$

$$z_4 = \cos\frac{4\pi}{3} + i\sin\frac{4\pi}{3} = -\frac{1}{2} - i\frac{\sqrt{3}}{2}$$

$$z_5 = \cos\frac{5\pi}{3} + i\sin\frac{5\pi}{3} = \frac{1}{2} - i\frac{\sqrt{3}}{2}$$

Notes

In this problem, the relations below have been used:

$$z = re^{i\theta} \Rightarrow \sqrt[n]{z} = \sqrt[n]{r}\left[\cos\left(\frac{2k\pi + \theta}{n}\right) + i\sin\left(\frac{2k\pi + \theta}{n}\right)\right], \quad k = 0, 1, 2, \ldots, n-1$$

$$\cos 0 = 1$$

$$\sin 0 = 0$$

$$\cos\frac{\pi}{3} = \frac{1}{2}$$

$$\sin\frac{\pi}{3} = \frac{\sqrt{3}}{2}$$

$$\cos\frac{2\pi}{3} = \cos\left(\pi - \frac{\pi}{3}\right) = -\cos\frac{\pi}{3} = -\frac{1}{2}$$

$$\sin\frac{2\pi}{3} = \sin\left(\pi - \frac{\pi}{3}\right) = \sin\frac{\pi}{3} = \frac{\sqrt{3}}{2}$$

$$\cos\pi = -1$$

$$\sin\pi = 0$$

$$\cos\frac{4\pi}{3} = \cos\left(\pi + \frac{\pi}{3}\right) = -\cos\frac{\pi}{3} = -\frac{1}{2}$$

$$\sin\frac{4\pi}{3} = \sin\left(\pi + \frac{\pi}{3}\right) = -\sin\frac{\pi}{3} = -\frac{\sqrt{3}}{2}$$

$$\cos\frac{5\pi}{3} = \cos\left(2\pi - \frac{\pi}{3}\right) = \cos\frac{\pi}{3} = \frac{1}{2}$$

$$\sin\frac{5\pi}{3} = \sin\left(2\pi - \frac{\pi}{3}\right) = -\sin\frac{\pi}{3} = -\frac{\sqrt{3}}{2}$$

1.4. Simplify the following complex quantity:

$$z = \left(\frac{1 + \sqrt{3}i}{1 - \sqrt{3}i} \right)^{10}$$

Difficulty level   ○ Easy   ● Normal   ○ Hard
Calculation amount   ● Small   ○ Normal   ○ Large

1) $-\dfrac{1}{2} + \dfrac{\sqrt{3}}{2}i$

2) $-2 - \sqrt{3}i$

3) $\dfrac{1}{2} - \dfrac{\sqrt{3}}{2}i$

4) $2 + \sqrt{3}i$

Partially Solved Exercise

Simplify the complex term below:

$$z = \frac{\left(1 + i\sqrt{3}\right)^8}{2^7\left(-1 + i\sqrt{3}\right)}$$

Solution

The complex quantities $1 + i\sqrt{3}$ and $-1 + i\sqrt{3}$ in polar form are as follows:

$$1 + i\sqrt{3} = (\quad)e^{i\overline{}}$$

$$-1 + i\sqrt{3} = (\quad)e^{i\overline{}}$$

Therefore:

$$z = \frac{\left((\quad)e^{i\overline{}}\right)^8}{2^7 \times (\quad)e^{i\overline{}}} = \frac{e^{i\overline{}}}{e^{i\overline{}}}$$

$$\Rightarrow z = e^{i(\quad)} = \cos(\quad) + i\sin(\quad)$$

$$\Rightarrow z = 1$$

Notes

In this problem, the relations below have been used:

$$z = a + ib = |z|e^{i\theta_z} \Rightarrow \begin{cases} |z| = \sqrt{a^2 + b^2},\, \theta_z = \tan^{-1}\left|\dfrac{b}{a}\right| & \text{if } a>0, b>0 \\[2mm] |z| = \sqrt{a^2 + b^2},\, \theta_z = \pi - \tan^{-1}\left|\dfrac{b}{a}\right| & \text{if } a<0, b>0 \\[2mm] |z| = \sqrt{a^2 + b^2},\, \theta_z = \pi + \tan^{-1}\left|\dfrac{b}{a}\right| & \text{if } a<0, b<0 \\[2mm] |z| = \sqrt{a^2 + b^2},\, \theta_z = -\tan^{-1}\left|\dfrac{b}{a}\right| & \text{if } a>0, b<0 \end{cases}$$

$$(e^a)^b = e^{ab}$$

$$\frac{e^a}{e^b} = e^{a-b}$$

$$e^{i\theta} = \cos\theta + i\sin\theta$$

$$\cos 2\pi = 1$$
$$\sin 2\pi = 0$$

1.5. Calculate the value of the term below:

$$z = \prod_{m=1}^{\infty} \left(\cos\frac{\pi}{2^m} + i\sin\frac{\pi}{2^m} \right)$$

Difficulty level   ○ Easy   ○ Normal   ● Hard
Calculation amount   ● Small   ○ Normal   ○ Large
1) -1
2) 1
3) πi
4) $\dfrac{\pi}{2}$

1.6. Calculate the value of the following term:

$$f(z) = \left| z e^{\frac{\pi}{3}i} - z \right|$$

Difficulty level   ○ Easy   ● Normal   ○ Hard
Calculation amount   ● Small   ○ Normal   ○ Large
1) $|z|$
2) $\dfrac{1}{2}|z|$
3) $\dfrac{1}{2}|z+1|$
4) $|z - i|$

Partially Solved Exercise

Calculate the value of the mathematical expression below:

$$z = \left[\left(\frac{e}{2} \right) \left(-1 - i\sqrt{3} \right) \right]^{3\pi i}$$

Solution

The complex quantities -1 and $1 + \sqrt{3}i$ in polar form are as follows:

$$-1 = (\quad) e^{i(\quad)}$$

$$1 + \sqrt{3}i = (\quad) e^{i(\quad)}$$

Thus:

$$z = \left[\left(\frac{e}{2} \right) (\quad) e^{i(\quad)} (\quad) e^{i(\quad)} \right]^{3\pi i}$$

$$\Rightarrow z = \left[(\quad) e^{i(\quad)} \right]^{3\pi i}$$

$$\Rightarrow z = -e^{-2\pi^2}$$

Notes

In this problem, the relations below have been used:

$$z = a + ib = |z| e^{i\theta_z} \Rightarrow \begin{cases} |z| = \sqrt{a^2 + b^2}, \theta_z = \tan^{-1}\left|\frac{b}{a}\right| & if \ a > 0, b > 0 \\[2mm] |z| = \sqrt{a^2 + b^2}, \theta_z = \pi - \tan^{-1}\left|\frac{b}{a}\right| & if \ a < 0, b > 0 \\[2mm] |z| = \sqrt{a^2 + b^2}, \theta_z = \pi + \tan^{-1}\left|\frac{b}{a}\right| & if \ a < 0, b < 0 \\[2mm] |z| = \sqrt{a^2 + b^2}, \theta_z = -\tan^{-1}\left|\frac{b}{a}\right| & if \ a > 0, b < 0 \end{cases}$$

$$(e^a)^b = e^{ab}$$

$$e^{i\theta} = \cos\theta + i\sin\theta$$

$$\cos 3\pi = \cos \pi = -1$$

$$\sin 3\pi = \sin \pi = 0$$

$$e^a e^b = e^{a+b}$$

1.7. Calculate the value of the following term:

$$z = \ln\left(\frac{-1 - i\sqrt{3}}{2}\right)$$

Difficulty level   ○ Easy   ● Normal   ○ Hard

Calculation amount   ● Small   ○ Normal   ○ Large

1) $-\frac{2\pi}{3}i$

2) $\frac{2\pi}{3}i$

3) $-\frac{\pi}{3}i$

4) $\ln\sqrt{3} + \frac{\pi}{3}i$

Exercise

Calculate the principal value of $\ln(-4)$.

Final Answer

$2\ln 2 + i\pi$

1.8. What is the principal value of $z = (1 - i)^{4i}$?

Difficulty level   ○ Easy   ○ Normal   ● Hard

Calculation amount   ● Small   ○ Normal   ○ Large

1) $e^{\pi + i2\ln 3}$

2) $e^{\pi + i4\ln 3}$

3) $e^{\pi + i2\ln 2}$

4) $e^{\pi + i3\ln 2}$

Partially Solved Exercise

Calculate the principal value of $z = (1 - i)^{1 + i}$.

Solution

The term can be written as follows:

$$z = e^{\ln(1-i)^{1+i}}$$

$$\Rightarrow z = e^{(1+i)(\qquad\qquad)}$$

$$\Rightarrow z = e^{(1+i)\left(\ln\left(\!\!\!\!\!\!\!\!\!\!\!\!\!\!\!\!\!\!\right)e^{}\right)\right)}$$

$$\Rightarrow z = e^{(1+i)(\ln()+())}$$

$$\Rightarrow z = e^{(()+i())}$$

$$\Rightarrow z = \sqrt{2}e^{()}e^{i()}$$

$$\Rightarrow z = \sqrt{2}e^{\frac{\pi}{4}}\left[\cos\left(\ln\sqrt{2} - \frac{\pi}{4}\right) + i\sin\left(\ln\sqrt{2} - \frac{\pi}{4}\right)\right]$$

Notes

In this problem, the relations below have been used:

$$z = e^{\ln z}$$

$$\ln z^a = a \ln z$$

$$\ln z = \ln\left|z\right|e^{i\theta_z} = \ln\left|z\right| + i\theta_z$$

$$z = a + ib = \left|z\right|e^{i\theta_z} \Rightarrow \begin{cases} \left|z\right| = \sqrt{a^2 + b^2}, \theta_z = \tan^{-1}\left|\dfrac{b}{a}\right| & \text{if } a > 0, b > 0 \\[2mm] \left|z\right| = \sqrt{a^2 + b^2}, \theta_z = \pi - \tan^{-1}\left|\dfrac{b}{a}\right| & \text{if } a < 0, b > 0 \\[2mm] \left|z\right| = \sqrt{a^2 + b^2}, \theta_z = \pi + \tan^{-1}\left|\dfrac{b}{a}\right| & \text{if } a < 0, b < 0 \\[2mm] \left|z\right| = \sqrt{a^2 + b^2}, \theta_z = -\tan^{-1}\left|\dfrac{b}{a}\right| & \text{if } a > 0, b < 0 \end{cases}$$

$$i^2 = -1$$

$$e^{a+b} = e^a e^b$$

$$e^{a+ib} = e^a e^{ib} = e^a(\cos b + i\sin b)$$

1.9. Calculate the value of the following function for $z = i$ where $\ln z$ is the principal branch of the natural logarithm.

$$f(z) = z^{\ln z}$$

Difficulty level   ○ Easy   ○ Normal   ● Hard
Calculation amount   ● Small   ○ Normal   ○ Large

1) $e^{-\frac{\pi^2}{4}}$

2) $e^{\frac{\pi^2}{4}}$

3) $e^{-\frac{\pi}{4}}$

4) $e^{-\pi^2}$

Partially Solved Exercise

If in $f(z) = z^{\ln z}$, $\ln z$ is the principal branch of the natural logarithm, calculate the value of $f(-1)$.

Solution

Based on the information given in the problem, we need to calculate the principal value of the following term:

$$f(-1) = (-1)^{\ln(-1)}$$

That term can be written as follows:

$$f(-1) = e^{\ln(-1)^{\ln(-1)}}$$

$$\Rightarrow f(-1) = e^{(\qquad \times \qquad)} \Rightarrow f(-1) = e^{(\qquad)^2}$$

$$\Rightarrow f(-1) = e^{\left(\ln(\quad)e^{\quad}\right)^2} \Rightarrow f(-1) = e^{\left(\ln(\quad) + (\quad)\right)^2}$$

$$\Rightarrow f(-1) = e^{(\qquad)^2}$$

$$\Rightarrow f(-1) = e^{-\pi^2}$$

Notes

In this problem, the relations below have been used:

$$z = e^{\ln z}$$

$$\ln z^a = a \ln z$$

$$\ln z = \ln|z|e^{i\theta_z} = \ln|z| + i\theta_z$$

$$z = a + ib = |z|e^{i\theta_z} \Rightarrow \begin{cases} |z| = \sqrt{a^2 + b^2}, \theta_z = \tan^{-1}\left|\dfrac{b}{a}\right| & \text{if } a>0, b>0 \\[2mm] |z| = \sqrt{a^2 + b^2}, \theta_z = \pi - \tan^{-1}\left|\dfrac{b}{a}\right| & \text{if } a<0, b>0 \\[2mm] |z| = \sqrt{a^2 + b^2}, \theta_z = \pi + \tan^{-1}\left|\dfrac{b}{a}\right| & \text{if } a<0, b<0 \\[2mm] |z| = \sqrt{a^2 + b^2}, \theta_z = -\tan^{-1}\left|\dfrac{b}{a}\right| & \text{if } a>0, b<0 \end{cases}$$

$$i^2 = -1$$

1.10. Calculate the principal value of i^{-i}.

Difficulty level   ○ Easy   ● Normal   ○ Hard
Calculation amount   ● Small   ○ Normal   ○ Large

1) $e^{-\frac{\pi}{2}}$

2) $e^{\frac{\pi}{2}}$

3) 1

4) -1

Partially Solved Exercise

Calculate the principal value of $(-1)^i$.

Solution

As we know:

$$-1 = (\quad)e^{i(\quad)}$$

Therefore:

$$(-1)^i = \left((\quad)e^{i(\quad)}\right)^i$$

$$\Rightarrow (-1)^i = e^{-\pi}$$

Notes

In this problem, the relations below have been used:

$$z = a + ib = |z|e^{i\theta_z} \Rightarrow \begin{cases} |z| = \sqrt{a^2 + b^2}, \theta_z = \tan^{-1}\left|\dfrac{b}{a}\right| & \text{if } a > 0, b > 0 \\[2mm] |z| = \sqrt{a^2 + b^2}, \theta_z = \pi - \tan^{-1}\left|\dfrac{b}{a}\right| & \text{if } a < 0, b > 0 \\[2mm] |z| = \sqrt{a^2 + b^2}, \theta_z = \pi + \tan^{-1}\left|\dfrac{b}{a}\right| & \text{if } a < 0, b < 0 \\[2mm] |z| = \sqrt{a^2 + b^2}, \theta_z = -\tan^{-1}\left|\dfrac{b}{a}\right| & \text{if } a > 0, b < 0 \end{cases}$$

$$(e^a)^b = e^{ab}$$

$$i^2 = -1$$

1.2 Limit of Complex Functions

1.11. Calculate the value of the following limit:

$$\lim_{z \to 0} \frac{xy}{x^2 + y^2}$$

Difficulty level   ● Easy   ○ Normal   ○ Hard
Calculation amount   ● Small   ○ Normal   ○ Large
1) 0
2) $\frac{1}{2}$
3) ∞
4) It does not exist.

Partially Solved Exercise

Calculate the value of the limit below:

$$\lim_{z \to 0} \left(\frac{x}{x+y} + ixy \right)$$

Solution

Regarding the first term, if we approach the origin through the path $y = mx$, we have:

$$\lim_{x \to 0} \frac{x}{x+y} = \lim_{x \to 0} \frac{x}{x + (\quad)} = \lim_{x \to 0} \frac{x}{x(\quad)} = \frac{1}{1+m}$$

Since the value of the limit of the first term depends on the path (the parameter m), the limit of the first term does not exist. Moreover, since the limit of the first term is unavailable, the total limit does not exist.

1.3 Complex Equations

1.12. Which one of the choices below is the expanded form of cos z?
Difficulty level   ○ Easy   ● Normal   ○ Hard
Calculation amount   ● Small   ○ Normal   ○ Large
1) $\cos^2 x \cosh y - i \sin^2 x \sinh y$
2) $\cos x \cosh^2 y - i \sin x \sinh^2 y$
3) $\cos x \sinh y - i \sin x \cosh y$
4) $\cos x \cosh y - i \sin x \sinh y$

1.13. Expand the term cos iz.
Difficulty level   ○ Easy   ● Normal   ○ Hard
Calculation amount   ● Small   ○ Normal   ○ Large

1) $\cos x \sinh y - i \sin x \cosh y$
2) $\cos x \cosh y - i \sin x \sinh y$
3) $\cos y \cosh x - i \sin y \sinh x$
4) $\cos y \cosh x + i \sin y \sinh x$

Exercise

Which one of the choices below is correct about $\cos iz$?
1) $\cos iz = \cosh z$
2) $\cos iz = i \cosh z$
3) $\cos iz = -i \cosh z$
4) $\cos iz = -\cosh z$

Final Answer
Choice (1)

1.14. Which one of the choices below is the expended form of $\sin z$?
   Difficulty level   ○ Easy   ● Normal   ○ Hard
   Calculation amount   ● Small   ○ Normal   ○ Large
   1) $\cos^2 x \cosh y - i \sin^2 x \sinh y$
   2) $\cos x \cosh y - i \sin x \sinh y$
   3) $\sin x \cosh y - i \sinh y \cos x$
   4) $\sin x \cosh y + i \sinh y \cos x$

1.15. Expand the term $\sin iz$.
   Difficulty level   ○ Easy   ● Normal   ○ Hard
   Calculation amount   ● Small   ○ Normal   ○ Large
   1) $-\sin y \cosh x - i \sinh x \cos y$
   2) $-\sin y \cosh x - i \sinh x \cos y$
   3) $\sin y \cosh x + i \sinh x \cos y$
   4) $-\sin y \cosh x + i \sinh x \cos y$

Exercise

Which one of the choices below is correct about $\sin iz$?
1) $\sin iz = i \sinh z$
2) $\sin iz = -i \sinh z$
3) $\sin iz = \sinh z$
4) $\sin iz = -\sinh z$

Final Answer
Choice (1)

1.16. Which one of the choices below is equal to $\sin z$?

Difficulty level   ● Easy   ○ Normal   ○ Hard
Calculation amount   ● Small   ○ Normal   ○ Large

1) $\dfrac{e^{iz} - e^{-iz}}{2i}$

2) $\dfrac{e^{iz} + e^{-iz}}{2}$

3) $\dfrac{e^{z} - e^{-z}}{2}$

4) $\dfrac{e^{z} + e^{-z}}{2}$

Exercise

Use Euler's formula to define $\sinh z$ based on exponential terms.

Final Answer

$\cosh z = \dfrac{e^{z} - e^{-z}}{2}$

1.17. Which one of the choices below is equal to $\cos z$?

Difficulty level   ● Easy   ○ Normal   ○ Hard
Calculation amount   ● Small   ○ Normal   ○ Large

1) $\dfrac{e^{iz} - e^{-iz}}{2i}$

2) $\dfrac{e^{iz} + e^{-iz}}{2}$

3) $\dfrac{e^{z} - e^{-z}}{2}$

4) $\dfrac{e^{z} + e^{-z}}{2}$

Exercise

Use Euler's formula to define $\cosh z$ based on exponential terms.

Final Answer

$\cosh z = \dfrac{e^{z} + e^{-z}}{2}$

1.18. Which one of the choices below is equal to $\tan z$?

Difficulty level   ○ Easy   ● Normal   ○ Hard
Calculation amount   ● Small   ○ Normal   ○ Large

1) $-i\dfrac{e^{2z} - 1}{e^{2z} + 1}$

2) $i\dfrac{e^{i2z} - 1}{e^{i2z} + 1}$

3) $\dfrac{e^{2z}-1}{e^{2z}+1}$

4) $-i\,\dfrac{e^{i2z}-1}{e^{i2z}+1}$

1.19. Which one of the choices below is equal to $\tanh z$?

Difficulty level   ○ Easy   ● Normal   ○ Hard
Calculation amount   ● Small   ○ Normal   ○ Large

1) $-i\,\dfrac{e^{2z}-1}{e^{2z}+1}$

2) $i\,\dfrac{e^{i2z}-1}{e^{i2z}+1}$

3) $\dfrac{e^{2z}-1}{e^{2z}+1}$

4) $-i\,\dfrac{e^{i2z}-1}{e^{i2z}+1}$

1.20. Solve the equation $\cosh z = 0$.

Difficulty level   ○ Easy   ● Normal   ○ Hard
Calculation amount   ○ Small   ● Normal   ○ Large

1) The equation does not have a solution.

2) $z_k = \left(k\pi - \dfrac{\pi}{2}\right)i,\ \forall k \epsilon \mathbb{Z}$

3) $z_k = \left(k\pi + \dfrac{\pi}{2}\right)i,\ \forall k \epsilon \mathbb{Z}$

4) $z_k = \left(2k\pi - \dfrac{\pi}{2}\right)i,\ \forall k \epsilon \mathbb{Z}$

Exercise

Solve the equation $\sinh z = 0$.

Final Answer

$z_k = ik\pi,\ \forall\ k\epsilon\mathbb{Z}$

1.21. Solve the following equation for $a > 0$.

$$Re\left(\frac{z-i}{z+i}\right) < 1, \quad Im\left(\frac{z-i}{z+i}\right) < a$$

Difficulty level   ○ Easy   ○ Normal   ● Hard
Calculation amount   ○ Small   ● Normal   ○ Large

1) The lower half and inner part of the circle $\left(x+\frac{1}{a}\right)^2 + (y+1)^2 = \frac{1}{a^2}$

2) The upper half and inner part of the circle $\left(x+\frac{1}{a}\right)^2 + (y+1)^2 = \frac{1}{a^2}$

3) The upper half and outer part of the circle $\left(x+\frac{1}{a}\right)^2 + (y+1)^2 = \frac{1}{a^2}$

4) The lower half and outer part of the circle $\left(x+\frac{1}{a}\right)^2 + (y+1)^2 = \frac{1}{a^2}$

Partially Solved Exercise

Solve the following equation for x:

$$a(\cos x + i \sin x) = 1 - i \tag{1}$$

$$a > 0 \tag{2}$$

Solution

From (1), it is concluded that:

$$\begin{cases} a \cos x = (\quad) \\ a \sin x = (\quad) \end{cases}, \tag{3}$$

$$\Rightarrow \tan x = (\quad) \tag{4}$$

$$\Rightarrow x = \qquad , \quad k = 0, \pm 1, \ldots \tag{5}$$

$$\Rightarrow x = \frac{3\pi}{4}, \frac{7\pi}{4}$$

$x = \frac{3\pi}{4}$ is not acceptable because the value of cosine and sine for that is negative and positive, respectively, which is inconsistent based on (2) and (3). Therefore:

$$x = \frac{7\pi}{4}$$

Notes

In this problem, the relations below have been used:

$$\tan x = \tan \alpha_0 \Rightarrow x = k\pi + \alpha_0, \quad k = 0, \pm 1, \ldots$$

1.22. Which one of the following choices is correct about the equation $e^z = -2$?
   Difficulty level   ○ Easy   ○ Normal   ● Hard
   Calculation amount   ○ Small   ● Normal   ○ Large
   1) The equation does not have any root.
   2) The equation has only one real root.
   3) The equation has only one imaginary root.
   4) The equation has an infinite number of roots.

1.23. Calculate the value of $z - \bar{z}$ if $|z + ai| = |z + bi|$ as well as the parameters a and b are real quantities where $a \neq b$.
   Difficulty level   ○ Easy   ● Normal   ○ Hard
   Calculation amount   ○ Small   ● Normal   ○ Large
   1) $-(a + b)i$
   2) $-(a - b)i$

3) $(a - b)i$
4) $(a + b)i$

1.24. Determine the region in which the value of the following term is purely real:

$$f(z) = z + \frac{1}{z}$$

Difficulty level   ○ Easy   ● Normal   ○ Hard
Calculation amount   ○ Small   ● Normal   ○ Large
1) The outer region of a circle centered at the origin with the radius 3 units.
2) A circle centered at the origin with the radius 2 units.
3) A circle centered at the origin with the radius 1 unit.
4) The inner region of a circle centered at the origin with the radius 1 unit.

Exercise

Determine the region in which the value of the following term is purely imaginary:

$$f(z) = z + \frac{1}{z}$$

Final Answer

y-axis

1.25. Define the x in terms of u and v in the following equation:

$$z = \frac{2}{u + iv}$$

Difficulty level   ○ Easy   ● Normal   ○ Hard
Calculation amount   ○ Small   ● Normal   ○ Large

1) $\dfrac{-v}{u^2 + v^2}$

2) $\dfrac{uv}{u^2 + v^2}$

3) $\dfrac{2u}{u^2 + v^2}$

4) $\dfrac{2u - v}{u^2 + v^2}$

Exercise

Define the y in terms of u and v in the following equation:

$$z = \frac{1}{u + iv}$$

Final Answer

$y = -\frac{v}{u^2 + v^2}$

1.26. Calculate the value of $|z_1 - z_2|$ where z_1 and z_2 are the roots of the equation $z^2 + z + 1 = i$.

Difficulty level   ○ Easy   ● Normal   ○ Hard

Calculation amount   ● Small   ○ Normal   ○ Large

1) $\sqrt{2}$

2) $\sqrt{3}$

3) $\sqrt{5}$

4) $\sqrt{10}$

Exercise

Calculate the value of $|z_1 - z_2|$ where z_1 and z_2 are the roots of the equation $10z^2 - 6z + 1 = 0$.

Final Answer

$\frac{1}{5}$

1.27. Which one of the regions below is the solution of the equation $|z - 1| \leq Im\ (z)$?

Difficulty level   ○ Easy   ● Normal   ○ Hard

Calculation amount   ● Small   ○ Normal   ○ Large

1) $x > 1$

2) $x = 1$

3) $y^2 \leq 2x - 1$

4) $y^2 \geq 2x - 1$

1.28. Solve the equation below:

$$\frac{2z}{1+i} - \frac{2z}{i} = \frac{5}{2+i}$$

Difficulty level   ● Easy   ○ Normal   ○ Hard

Calculation amount   ○ Small   ● Normal   ○ Large

1) $\frac{1}{2}(1 - 3i)$

2) $\frac{1}{2}(1 + 3i)$

3) $\frac{1}{2}(-1 - 3i)$

4) $\frac{1}{2}(-1 + 3i)$

1.29. What is the shape of solution of the equation $z\bar{z} = 36$?

Difficulty level   ● Easy   ○ Normal   ○ Hard

Calculation amount   ● Small   ○ Normal   ○ Large

1) A circle

2) A hyperbola

3) An ellipse

4) A parabola

Exercise

Determine the region that indicates the solution of the equation below:

$$\bar{z} = \frac{1}{z}$$

Final Answer

A unit circle centered at the origin

1.30. Determine the equation of a circle with radius four centered at $(-2, 1)$.
   Difficulty level   ● Easy   ○ Normal   ○ Hard
   Calculation amount   ● Small   ○ Normal   ○ Large
   1) $|z + 2 - i| = 4$
   2) $|z + 2 + i| = 4$
   3) $|z - 2 + i| = 4$
   4) $|z - 2 - i| = 4$

Exercise

Determine the equation of a unit circle centered at $(1, 1)$.

Final Answer

$|z - 1 - i| = 1$

1.31. What is the shape of solution of the equation below?

$$\left|\frac{z - 3i}{z + i}\right| = 1$$

   Difficulty level   ● Easy   ○ Normal   ○ Hard
   Calculation amount   ○ Small   ● Normal   ○ Large
   1) An ellipse
   2) The line $y = 0$
   3) The lines $y = -x$ and $y = x$
   4) The line $y = 1$

1.32. What is the shape of solution of the equation $|z - 2| = |z + 4|$?
   Difficulty level   ● Easy   ○ Normal   ○ Hard
   Calculation amount   ○ Small   ● Normal   ○ Large
   1) A circle
   2) An ellipse
   3) A parabola
   4) A line

1.33. What is the shape of solution of the equation $|z|^2 + 3Re(z^2) = 4$?

Difficulty level   ○ Easy   ● Normal ○ Hard

Calculation amount   ○ Small ● Normal ○ Large

1) A hyperbola
2) An ellipse
3) A parabola
4) A circle

Exercise

Determine the shape of solution of the equation $|z|^2 + Re\,(z^2) = 0$.

Final Answer

A line

1.34. What is the shape of solution of the equation below?

$$\left|\frac{z+i}{z-i}\right| = \sqrt{2}$$

Difficulty level   ● Easy   ○ Normal ○ Hard

Calculation amount   ○ Small ● Normal ○ Large

1) The line $x = y$
2) A circle with the radius $\sqrt{2}$ and the center at $(3, 0)$
3) A circle with the radius $2\sqrt{2}$ and the center at $(0, 3)$
4) A circle with the radius $2\sqrt{2}$ and the center at $(1, 1)$

1.35. Which of the following choices is correct about the term below?

$$\left|\frac{z_1 - \bar{z}_2}{z_1 + \bar{z}_2}\right| = 1$$

Difficulty level   ○ Easy   ● Normal ○ Hard

Calculation amount   ○ Small ● Normal ○ Large

1) $Re(z_1 z_2) > 0$
2) $Re(z_1 z_2) < 0$
3) $Re(z_1 z_2) = 0$
4) $Im(z_1 z_2) > 0$

1.36. Calculate the value of $u(x, y)$ and $v(x, y)$ from the following equation:

$$u(x, y) + iv(x, y) = \frac{1}{1 - z}$$

Difficulty level   ○ Easy   ● Normal ○ Hard

Calculation amount   ○ Small ● Normal ○ Large

1) $u(x, y) = \dfrac{1 - x}{1 + x}, \quad v(x, y) = \dfrac{1 - y}{1 - x}$

2) $u(x, y) = \dfrac{y}{1 - x}, \quad v(x, y) = \dfrac{x}{1 - y}$

3) $u(x, y) = \dfrac{y}{(1-x)^2 + y^2}$, $v(x, y) = \dfrac{1-x}{(1-x)^2 + y^2}$

4) $u(x, y) = \dfrac{1-x}{(1-x)^2 + y^2}$, $v(x, y) = \dfrac{y}{(1-x)^2 + y^2}$

1.4   Holomorphic (Analytic) Function and Its Harmonic Conjugate Functions

1.37. Determine the harmonic conjugate function of $u(x, y) = 2x(3 - y)$ if $f(x + jy) = u(x, y) + jv(x, y)$ is a complex holomorphic function.

Difficulty level   ○ Easy   ● Normal   ○ Hard
Calculation amount   ○ Small   ○ Normal   ● Large

1) $2x(3 + y) + c$
2) $x^2 - y^2 + c$
3) $-2x(3 + y) + c$
4) $x^2 - (3 - y)^2 + c$

Partially Solved Exercise

Assume that $f(x + jy) = u(x, y) + jv(x, y)$ is a complex holomorphic function. Calculate the harmonic conjugate function of $u(x, y) = e^{-x}(x \sin y - y \cos y)$.

Solution

Herein, $u(x, y)$ and its harmonic conjugate function, that is, $v(x, y)$, must satisfy the Cauchy-Riemann equations. In other words:

$$u_x(x, y) = v_y(x, y) \tag{1}$$

$$u_y(x, y) = -v_x(x, y) \tag{2}$$

Based on the information given in the problem, we have:

$$u(x, y) = e^{-x}(x \sin y - y \cos y) \tag{3}$$

Solving (1) and (3):

$$\frac{\partial}{\partial x}\left(e^{-x}(x \sin y - y \cos y)\right) = v_y(x, y) \tag{4}$$

$$\Rightarrow v_y(x, y) = (\qquad\qquad) \tag{5}$$

$$\Rightarrow v(x, y) = \int (\qquad\qquad)dy$$

$$\Rightarrow v(x, y) = (\qquad\qquad) + h(x) \tag{6}$$

On the other hand, by solving (2), (3), and (6), we have:

$$\frac{\partial}{\partial y}(e^{-x}(x\sin y - y\cos y)) = -\frac{\partial}{\partial x}[(\qquad\qquad) + h(x)]$$

$$(\qquad\qquad) = -(\qquad\qquad) - h'(x) \tag{7}$$

$$\Rightarrow h'(x) = (\qquad) \tag{8}$$

$$\Rightarrow h(x) = (\qquad) \tag{9}$$

Solving (6) and (9):

$$v(x, y) = e^{-x}(y\sin y + x\cos y) + c$$

Exercise

Calculate the harmonic conjugate function of $u(x, y) = y^3 - 3x^2y$. Herein, assume that $f(x + jy) = u(x, y) + jv(x, y)$ is a complex holomorphic function.

Final Answer

$-3xy^2 + x^3 + c$

1.38. Determine the harmonic conjugate function of $u(x, y) = \ln(x^2 + y^2)$.
   Difficulty level   ○ Easy   ● Normal   ○ Hard
   Calculation amount   ○ Small   ○ Normal   ● Large
   1) $v(x, y) = \frac{1}{2}\cot^{-1}\frac{x}{y} + c$
   2) $v(x, y) = \frac{1}{2}\tan^{-1}\frac{x}{y} + c$
   3) $v(x, y) = 2\tan^{-1}\frac{y}{x} + c$
   4) $v(x, y) = 2\cot^{-1}\frac{y}{x} + c$

Partially Solved Exercise

Calculate the harmonic conjugate function of $u(x, y) = x^2 - y^2 + 2x$.

Solution

Based on the information given in the problem, we have:

$$u(x, y) = x^2 - y^2 + 2x \tag{1}$$

Herein, $u(x, y)$ and its harmonic conjugate function, that is, $v(x, y)$, must satisfy the Cauchy-Riemann equations. In other words:

$$u_x(x, y) = v_y(x, y) \tag{2}$$

$$u_y(x, y) = -v_x(x, y) \tag{3}$$

Solving (1) and (2):

$$\frac{\partial}{\partial x}\left(x^2 - y^2 + 2x\right) = v_y(x, y) \tag{4}$$

$$\Rightarrow v_y(x, y) = (\qquad) \tag{5}$$

$$\Rightarrow v(x, y) = \int (\qquad) dy$$

$$\Rightarrow v(x, y) = \qquad + h(x) \tag{6}$$

On the other hand, by solving (1), (3), and (6), we have:

$$\frac{\partial}{\partial y}\left(x^2 - y^2 + 2x\right) = -\frac{\partial}{\partial x}[(\qquad) + h(x)]$$

$$(\qquad) = -(\qquad) - h'(x) \tag{7}$$

$$\Rightarrow h'(x) =$$

$$\Rightarrow h(x) = \tag{8}$$

Solving (6) and (8):

$$v(x, y) = (2x + 2)y + c$$

Exercise

Find the harmonic conjugate function of $u(x, y) = 2^x \cos (y \ln 2)$.

Final Answer

$v(x, y) = 2^x \sin (y \ln 2) + c$

1.39. For what value of a and b, the function $u(x, y) = x^2 + ay^2 + bxy$ is a harmonic function.
    Difficulty level   ○ Easy   ● Normal   ○ Hard
    Calculation amount   ● Small   ○ Normal   ○ Large
    1) $a = b = 1$
    2) $a = -1, b \epsilon \mathbb{R}$
    3) $a = -1, b = 0$
    4) For no value of a and b

1.40. Calculate the value of the following term if $u(x, y)$ is a harmonic function:

$$\frac{\partial^2 u(x, y)}{\partial z \partial \bar{z}}$$

Difficulty level   ○ Easy   ○ Normal   ● Hard
Calculation amount   ○ Small   ○ Normal   ● Large

1) 0

2) $\dfrac{1}{4}$

3) $\dfrac{1}{4i}$

4) $-\dfrac{1}{4}$

1.41. For what value of α and β, the function $u(x, y) = x^3 + \alpha x^2 y + \beta x y^2 + y^3$ is a harmonic function?
Difficulty level   ○ Easy   ● Normal   ○ Hard
Calculation amount   ○ Small   ● Normal   ○ Large

1) $\alpha = \beta = -3$
2) $\alpha = \beta = -2$
3) $\alpha = -\beta = 3$
4) $\alpha = -\beta = -3$

1.42. Assume that $f(z)$ is a harmonic function where its real part is $u(x, y) = x + e^x \cos y$. Calculate the value of $f'(1)$.
Difficulty level   ○ Easy   ● Normal   ○ Hard
Calculation amount   ○ Small   ● Normal   ○ Large

1) $1 + e$
2) $1 - e$
3) $e + 2i$
4) $(1 + e) + i$

Exercise

$u(x, y) = x^2 + \ln y$ is the real part of the harmonic function $f(z)$. Calculate the value of $f'(1 - i)$.

Final Answer

$2 + i$

References

1. Rahmani-Andebili, M. (2024). Precalculus (2nd Ed.) – Practice Problems, Methods, and Solutions, Springer Nature.
2. Rahmani-Andebili, M. (2023). Calculus III – Practice Problems, Methods, and Solutions, Springer Nature.
3. Rahmani-Andebili, M. (2023). Calculus II – Practice Problems, Methods, and Solutions, Springer Nature.
4. Rahmani-Andebili, M. (2023). Calculus I (2nd Ed.) – Practice Problems, Methods, and Solutions, Springer Nature.
5. Rahmani-Andebili, M. (2022). Differential Equations – Practice Problems, Methods, and Solutions, Springer Nature.
6. Rahmani-Andebili, M. (2021). Calculus – Practice Problems, Methods, and Solutions, Springer Nature.
7. Rahmani-Andebili, M. (2021). Precalculus – Practice Problems, Methods, and Solutions, Springer Nature.

Complex Functions, Equations, Quantities, and Limits: Solutions of Problems

2

Abstract

In this chapter, the problems of the first chapter are fully solved, in detail, step-by-step, and with different methods.

2.1 Complex Quantities

2.1. Based on the information given in the problem, we need to calculate the term $f(z) = (x^2 - y^2) + i(x + y^2)$ for $z = 1 + 2i$ [1–7].

From $z = x + iy = 1 + 2i$, it is noticed that $x = 1$ and $y = 2$. Thus:

$$f(1 + 2i) = (1^2 - 2^2) + i(1 + 2^2)$$

$$\Rightarrow f(1 + 2i) = -3 + 5i$$

Choice (3) is the answer.

2.2. Based on the information given in the problem, the value of $z^{100}(1 - i)$ needs to be calculated for $z = \frac{\sqrt{2}}{2} + i\frac{\sqrt{2}}{2}$.

The value of $z = \frac{\sqrt{2}}{2} + i\frac{\sqrt{2}}{2}$ in polar form is as follows:

$$z = e^{i\frac{\pi}{4}}$$

Thus:

$$z^{100}(1 - i) = \left(e^{i\frac{\pi}{4}}\right)^{100}(1 - i)$$

$$= \left(e^{25\pi i}\right)(1 - i) = \left(e^{\pi i}\right)(1 - i)$$

$$= (\cos \pi + i \sin \pi)(1 - i) = -(1 - i)$$

$$= -1 + i$$

Choice (3) is the answer.

Notes

In this problem, the relations below have been used:

$$z = a + ib = |z|e^{i\theta_z} \Rightarrow \begin{cases} |z| = \sqrt{a^2 + b^2}, \theta_z = \tan^{-1}\left|\dfrac{b}{a}\right| & \text{if } a > 0, b > 0 \\[2mm] |z| = \sqrt{a^2 + b^2}, \theta_z = \pi - \tan^{-1}\left|\dfrac{b}{a}\right| & \text{if } a < 0, b > 0 \\[2mm] |z| = \sqrt{a^2 + b^2}, \theta_z = \pi + \tan^{-1}\left|\dfrac{b}{a}\right| & \text{if } a < 0, b < 0 \\[2mm] |z| = \sqrt{a^2 + b^2}, \theta_z = -\tan^{-1}\left|\dfrac{b}{a}\right| & \text{if } a > 0, b < 0 \end{cases}$$

$$(z)^n = \left(|z|e^{i\theta_z}\right)^n = |z|^n e^{in\theta_z}$$

$$e^{i\theta} = \cos\theta + i\sin\theta$$

$$\cos\pi = -1$$

$$\sin\pi = 0$$

2.3. Based on the information given in the problem, we need to calculate one of the fourth roots of $z = -9i$.

This complex quantity can be presented in polar form as follows:

$$z = 9e^{-\frac{\pi}{2}i}$$

Now, the fourth root of the complex quantity can be calculated as follows:

$$\sqrt[4]{z} = \sqrt[4]{9}\left[\cos\left(\frac{2k\pi - \frac{\pi}{2}}{4}\right) + i\sin\left(\frac{2k\pi - \frac{\pi}{2}}{4}\right)\right], \quad k = 0, 1, 2, 3$$

$$\Rightarrow \sqrt[4]{z} = \sqrt{3}\left[\cos\left(\frac{k\pi}{2} - \frac{\pi}{8}\right) + i\sin\left(\frac{k\pi}{2} - \frac{\pi}{8}\right)\right], \quad k = 0, 1, 2, 3$$

For $k = 3$, we have:

$$\sqrt[4]{z} = \sqrt{3}\left[\cos\frac{11\pi}{8} + i\sin\left(\frac{11\pi}{8}\right)\right]$$

Choice (3) is the answer.

Notes

In this problem, the relation below has been used:

$$z = re^{i\theta} \Rightarrow \sqrt[n]{z} = \sqrt[n]{r}\left[\cos\left(\frac{2k\pi + \theta}{n}\right) + i\sin\left(\frac{2k\pi + \theta}{n}\right)\right], \quad k = 0, 1, 2, \ldots, n-1$$

2.4. Based on the information given in the problem, we have:

$$z = \left(\frac{1 + \sqrt{3}i}{1 - \sqrt{3}i}\right)^{10}$$

This complex quantity can be presented in polar form as follows:

$$z = \left(\frac{2e^{i\frac{\pi}{3}}}{2e^{-i\frac{\pi}{3}}}\right)^{10} = \left(e^{i\frac{2\pi}{3}}\right)^{10} = e^{i\frac{20\pi}{3}} = e^{i\frac{2\pi}{3}}$$

$$\Rightarrow z = \cos\frac{2\pi}{3} + i\sin\frac{2\pi}{3} \Rightarrow z = -\frac{1}{2} + \frac{\sqrt{3}}{2}i$$

Choice (1) is the answer.

Notes

In this problem, the relations below have been used:

$$z = a + ib = |z|e^{i\theta_z} \Rightarrow \begin{cases} |z| = \sqrt{a^2 + b^2}, \theta_z = \tan^{-1}\left|\frac{b}{a}\right| & \text{if } a > 0, b > 0 \\ |z| = \sqrt{a^2 + b^2}, \theta_z = \pi - \tan^{-1}\left|\frac{b}{a}\right| & \text{if } a < 0, b > 0 \\ |z| = \sqrt{a^2 + b^2}, \theta_z = \pi + \tan^{-1}\left|\frac{b}{a}\right| & \text{if } a < 0, b < 0 \\ |z| = \sqrt{a^2 + b^2}, \theta_z = -\tan^{-1}\left|\frac{b}{a}\right| & \text{if } a > 0, b < 0 \end{cases}$$

$$\frac{e^a}{e^b} = e^{a-b}$$

$$(e^a)^b = e^{ab}$$

$$e^{i\theta} = \cos\theta + i\sin\theta$$

$$\cos\frac{2\pi}{3} = -\cos\frac{\pi}{3} = -\frac{1}{2}$$

$$\sin\frac{2\pi}{3} = \sin\frac{\pi}{3} = \frac{\sqrt{3}}{2}$$

2.5. Based on the information given in the problem, we have:

$$z = \prod_{m=1}^{\infty}\left(\cos\frac{\pi}{2^m} + i\sin\frac{\pi}{2^m}\right)$$

The problem can be solved as follows:

$$z = \prod_{m=1}^{\infty} e^{i\frac{\pi}{2^m}} = e^{i\frac{\pi}{2^1}} e^{i\frac{\pi}{2^2}} e^{i\frac{\pi}{2^3}} e^{i\frac{\pi}{2^\infty}} = e^{i\pi\left(\frac{1}{2} + \frac{1}{2^2} + \frac{1}{2^3} + \cdots + \frac{1}{2^\infty}\right)} \tag{1}$$

As we know, the limit sum of a geometric sequence can be calculated as follows:

$$S = \frac{a_1}{1-q}$$

Therefore:

$$\frac{1}{2} + \frac{1}{2^2} + \frac{1}{2^3} + + \frac{1}{2^\infty} = \frac{\frac{1}{2}}{1-\frac{1}{2}} = 1 \tag{2}$$

Solving (1) and (2):

$$z = e^{\pi i} = \cos \pi + i \sin \pi \Rightarrow z = -1$$

Choice (1) is the answer.

Notes

In this problem, the relations below have been used:

$$e^{i\theta} = \cos \theta + i \sin \theta$$

$$e^a e^b = e^{a+b}$$

$$a_1 + a_1 q^1 + a_1 q^2 + + a_1 q^\infty = \frac{a_1}{1-q}$$

$$\cos \pi = -1$$

$$\sin \pi = 0$$

2.6. Based on the information given in the problem, we have:

$$f(z) = \left| z e^{\frac{\pi}{3}i} - z \right|$$

The problem can be solved as follows:

$$f(z) = \left| z \left(e^{\frac{\pi}{3}i} - 1 \right) \right| = |z| \left| e^{\frac{\pi}{3}i} - 1 \right| = |z| \left| \cos \frac{\pi}{3} + i \sin \frac{\pi}{3} - 1 \right|$$

$$\Rightarrow f(z) = |z| \left| \left(\frac{1}{2} - 1 \right) + i \frac{\sqrt{3}}{2} \right| = |z| \left| -\frac{1}{2} + i \frac{\sqrt{3}}{2} \right| = |z| \sqrt{\left(-\frac{1}{2} \right)^2 + \left(\frac{\sqrt{3}}{2} \right)^2}$$

$$\Rightarrow f(z) = |z|$$

Choice (1) is the answer.

Notes

In this problem, the relations below have been used:

$$|f(z)g(z)| = |f(z)||g(z)|$$

$$e^{i\theta} = \cos\theta + i\sin\theta$$

$$|a + ib| = \sqrt{a^2 + b^2}$$

2.7. Based on the information given in the problem, we have:

$$z = \ln\left(\frac{-1 + i\sqrt{3}}{2}\right)$$

The problem can be solved as follows:

$$z = \ln\left(-\frac{1}{2} + i\frac{\sqrt{3}}{2}\right)$$

$$\Rightarrow z = \ln 1 e^{\frac{2\pi}{3}i} = \ln 1 + \ln e^{\frac{2\pi}{3}i}$$

$$\Rightarrow z = \frac{2\pi}{3}i$$

Choice (2) is the answer.

Notes

In this problem, the relations below have been used:

$$z = a + ib = |z|e^{i\theta_z} \Rightarrow \begin{cases} |z| = \sqrt{a^2 + b^2}, \theta_z = \tan^{-1}\left|\frac{b}{a}\right| & if \ a > 0, b > 0 \\ |z| = \sqrt{a^2 + b^2}, \theta_z = \pi - \tan^{-1}\left|\frac{b}{a}\right| & if \ a < 0, b > 0 \\ |z| = \sqrt{a^2 + b^2}, \theta_z = \pi + \tan^{-1}\left|\frac{b}{a}\right| & if \ a < 0, b < 0 \\ |z| = \sqrt{a^2 + b^2}, \theta_z = -\tan^{-1}\left|\frac{b}{a}\right| & if \ a > 0, b < 0 \end{cases}$$

$$\ln|z|e^{i\theta_z} = \ln|z| + i\theta_z$$

2.8. Based on the information given in the problem, we need to calculate the principal value of the following term:

$$z = (1 - i)^{4i}$$

The term can be written as follows:

$$z = e^{\ln(1-i)^{4i}}$$

$$\Rightarrow z = e^{4i\ln(1-i)} \Rightarrow z = e^{4i\left(\ln\sqrt{2}e^{-i\frac{\pi}{4}}\right)}$$

$$\Rightarrow z = e^{4i\left(\ln\sqrt{2}-i\frac{\pi}{4}\right)} = e^{i4\ln\sqrt{2}+\pi} \Rightarrow z = e^{\pi+i2\ln 2}$$

Choice (3) is the answer.

Notes

In this problem, the relations below have been used:

$$z = e^{\ln z}$$

$$\ln z^a = a\ln z$$

$$\ln z = \ln\left|z\right|e^{i\theta_z} = \ln\left|z\right| + i\theta_z$$

$$z = a + ib = \left|z\right|e^{i\theta_z} \Rightarrow \begin{cases} \left|z\right| = \sqrt{a^2 + b^2}, \theta_z = \tan^{-1}\left|\dfrac{b}{a}\right| & \text{if } a > 0, b > 0 \\[2mm] \left|z\right| = \sqrt{a^2 + b^2}, \theta_z = \pi - \tan^{-1}\left|\dfrac{b}{a}\right| & \text{if } a < 0, b > 0 \\[2mm] \left|z\right| = \sqrt{a^2 + b^2}, \theta_z = \pi + \tan^{-1}\left|\dfrac{b}{a}\right| & \text{if } a < 0, b < 0 \\[2mm] \left|z\right| = \sqrt{a^2 + b^2}, \theta_z = -\tan^{-1}\left|\dfrac{b}{a}\right| & \text{if } a > 0, b < 0 \end{cases}$$

$$i^2 = -1$$

2.9. Based on the information given in the problem, we have:

$$f(z) = z^{\ln z}$$

$$\Rightarrow \ln f(z) = \ln\left(z^{\ln z}\right)$$

$$\Rightarrow \ln f(z) = \ln z \ln z = (\ln z)^2$$

$$\Rightarrow f(z) = e^{(\ln z)^2}$$

For $z = i$, we have:

$$f(i) = e^{(\ln i)^2}$$

$$\Rightarrow f(i) = e^{\left(\ln 1e^{\frac{\pi}{2}i}\right)^2} = e^{\left(\ln 1+\frac{\pi}{2}i\right)^2} = e^{\left(0+\frac{\pi}{2}i\right)^2}$$

$$\Rightarrow f(i) = e^{-\frac{\pi^2}{4}}$$

Choice (1) is the answer.

Notes

In this problem, the relations below have been used:

$$\ln z^a = a \ln z$$

$$\ln a = b \Rightarrow a = e^b$$

$$\ln z = \ln|z|e^{i\theta_z} = \ln|z| + i\theta_z$$

$$z = a + ib = |z|e^{i\theta_z} \Rightarrow \begin{cases} |z| = \sqrt{a^2 + b^2}, \theta_z = \tan^{-1}\left|\dfrac{b}{a}\right| & if \ a>0, b>0 \\[2mm] |z| = \sqrt{a^2 + b^2}, \theta_z = \pi - \tan^{-1}\left|\dfrac{b}{a}\right| & if \ a<0, b>0 \\[2mm] |z| = \sqrt{a^2 + b^2}, \theta_z = \pi + \tan^{-1}\left|\dfrac{b}{a}\right| & if \ a<0, b<0 \\[2mm] |z| = \sqrt{a^2 + b^2}, \theta_z = -\tan^{-1}\left|\dfrac{b}{a}\right| & if \ a>0, b<0 \end{cases}$$

$$\ln 1 = 0$$

$$i^2 = -1$$

2.10. Based on the information given in the problem, the value of i^{-i} needs to be calculated.

As we know:

$$i = 1e^{\frac{\pi}{2}i} = e^{\frac{\pi}{2}i}$$

Hence:

$$i^{-i} = \left(e^{\frac{\pi}{2}i}\right)^{-i} = e^{\frac{\pi}{2}}$$

Choice (2) is the answer.

Notes

In this problem, the relations below have been used:

$$z = a + ib = |z|e^{i\theta_z} \Rightarrow \begin{cases} |z| = \sqrt{a^2 + b^2}, \theta_z = \tan^{-1}\left|\dfrac{b}{a}\right| & if \ a>0, b>0 \\[2mm] |z| = \sqrt{a^2 + b^2}, \theta_z = \pi - \tan^{-1}\left|\dfrac{b}{a}\right| & if \ a<0, b>0 \\[2mm] |z| = \sqrt{a^2 + b^2}, \theta_z = \pi + \tan^{-1}\left|\dfrac{b}{a}\right| & if \ a<0, b<0 \\[2mm] |z| = \sqrt{a^2 + b^2}, \theta_z = -\tan^{-1}\left|\dfrac{b}{a}\right| & if \ a>0, b<0 \end{cases}$$

$$\left(e^a\right)^b = e^{ab} \qquad \cdot$$

$$i^2 = -1$$

2.2   Limit of Complex Functions

2.11. Based on the information given in the problem, we have:

$$\lim_{z \to 0} \frac{xy}{x^2 + y^2}$$

If we approach the origin through the path $y = mx$, we have:

$$\lim_{x \to 0} \frac{x(mx)}{x^2 + (mx)^2} = \lim_{x \to 0} \frac{m}{1 + m^2} = \frac{m}{1 + m^2}$$

Since the value of the limit depends on the path (the parameter m), the limit does not exist. Choice (4) is the answer.

2.3   Complex Equations

2.12. Based on the information given in the problem, we need to extend the value of cos$z$.

As we know:

$$z = x + iy$$

Thus:

$$\cos z = \cos(x + iy) = \cos(x)\cos(iy) - \sin(x)\sin(iy)$$

$$\Rightarrow \cos z = \cos x \cosh y - i \sin x \sinh y$$

Choice (4) is the answer.

> **Notes**
>
> In this problem, the relations below have been used:
>
> $$\cos(a + b) = \cos(a)\cos(b) - \sin(a)\sin(b)$$
>
> $$\cos(ix) = \cosh x$$
>
> $$\sin(ix) = i \sinh x$$

2.13. Based on the information given in the problem, we need to extend the value of cos iz.

As we know:

$$z = x + iy$$

Thus:

$$\cos iz = \cos i(x + iy) = \cos(-y + ix)$$

$$\Rightarrow \ \cos iz = \cos(-y)\cos(ix) - \sin(-y)\sin(ix)$$

$$\Rightarrow \ \cos iz = \cos y \cosh x + i \sin y \sinh x$$

Choice (4) is the answer.

Notes

In this problem, the relations below have been used:

$$\cos(a+b) = \cos(a)\cos(b) - \sin(a)\sin(b)$$

$$\cos(-x) = \cos x$$

$$\cos(ix) = \cosh x$$

$$\sin(-x) = -\sin x$$

$$\sin(ix) = i \sinh x$$

$$i^2 = -1$$

2.14. Based on the information given in the problem, we need to extend the value of sin z.

As we know:

$$z = x + iy$$

Thus:

$$\sin z = \sin(x+iy) = \sin(x)\cos(iy) + \sin(iy)\cos(x)$$

$$\Rightarrow \ \sin z = \sin x \cosh y + i \sinh y \cos x$$

Choice (4) is the answer.

Notes

In this problem, the relations below have been used:

$$\sin(a+b) = \sin(a)\cos(b) + \sin(b)\cos(a)$$

$$\cos(ix) = \cosh x$$

$$\sin(ix) = i \sinh x$$

2.15. Based on the information given in the problem, we need to extend the value of sin iz.

As we know:

$$z = x + iy$$

Thus:

$$\sin iz = \sin i(x + iy) = \sin(-y + ix)$$

$$\Rightarrow \sin iz = \sin(-y)\cos(ix) + \sin(ix)\cos(-y)$$

$$\Rightarrow \sin iz = -\sin y \cosh x + i \sinh x \cos y$$

Choice (4) is the answer.

Notes

In this problem, the relations below have been used:

$$\sin(a + b) = \sin(a)\cos(b) + \sin(b)\cos(a)$$

$$\sin(-x) = -\sin x$$

$$\cos(ix) = \cosh x$$

$$\sin(ix) = i \sinh x$$

$$\cos(-x) = \cos x$$

$$i^2 = -1$$

2.16. Based on Euler's formula, we have:

$$e^{iz} = \cos z + i \sin z \tag{1}$$

Applying Euler's formula for $-iz$:

$$e^{-iz} = \cos z - i \sin z \tag{2}$$

Subtracting (2) from (1) results in the following relation:

$$\sin z = \frac{e^{iz} - e^{-iz}}{2i}$$

Choice (1) is the answer.

2.17. Based on Euler's formula, we have:

$$e^{iz} = \cos z + i \sin z \tag{1}$$

Applying Euler's formula for $-iz$:

$$e^{-iz} = \cos z - i \sin z \tag{2}$$

Adding (2) to (1) results in the following relation:

$$\cos z = \frac{e^{iz} + e^{-iz}}{2}$$

Choice (2) is the answer.

2.18. As we know:

$$\tan z = \frac{\sin z}{\cos z}$$

$$\sin z = \frac{e^{iz} - e^{-iz}}{2i}$$

$$\cos z = \frac{e^{iz} + e^{-iz}}{2}$$

Thus:

$$\tan z = \frac{\frac{e^{iz} - e^{-iz}}{2i}}{\frac{e^{iz} + e^{-iz}}{2}} = \frac{e^{iz} - e^{-iz}}{i(e^{iz} + e^{-iz})}$$

$$\Rightarrow \tan z = \frac{e^{i2z} - 1}{i(e^{i2z} + 1)} \Rightarrow \tan z = -i\frac{e^{i2z} - 1}{e^{i2z} + 1}$$

Choice (4) is the answer.

Notes

In this problem, the relation below has been used:

$$i^2 = -1$$

2.19. As we know:

$$\tanh z = \frac{\sinh z}{\cosh z}$$

$$\sinh z = \frac{e^z - e^{-z}}{2}$$

$$\cosh z = \frac{e^z + e^{-z}}{2}$$

Thus:

$$\tanh z = \frac{\frac{e^z - e^{-z}}{2}}{\frac{e^z + e^{-z}}{2}} = \frac{e^z - e^{-z}}{e^z + e^{-z}}$$

$$\Rightarrow \tanh z = \frac{e^{2z} - 1}{e^{2z} + 1}$$

Choice (3) is the answer.

2.20. Based on the information given in the problem, we have:

$$\cosh z = 0$$

As we know:

$$\cosh z = \frac{e^z + e^{-z}}{2}$$

Therefore:

$$\frac{e^z + e^{-z}}{2} = 0 \Rightarrow e^z + e^{-z} = 0$$

$$\Rightarrow e^{2z} + 1 = 0 \Rightarrow e^{2z} = -1$$

$$\Rightarrow e^{2z} = 1e^{i(2k\pi + \pi)} \Rightarrow 2z_k = i(2k\pi + \pi), \forall k \epsilon \mathbb{Z}$$

$$\Rightarrow z_k = \left(k\pi + \frac{\pi}{2}\right)i, \forall k \epsilon \mathbb{Z}$$

Choice (3) is the answer.

Notes

In this problem, the relations below have been used:

$$z = a + ib = |z|e^{i\theta_z} \Rightarrow \begin{cases} |z| = \sqrt{a^2 + b^2}, \theta_z = \tan^{-1}\left|\frac{b}{a}\right| & if \ a > 0, b > 0 \\ |z| = \sqrt{a^2 + b^2}, \theta_z = \pi - \tan^{-1}\left|\frac{b}{a}\right| & if \ a < 0, b > 0 \\ |z| = \sqrt{a^2 + b^2}, \theta_z = \pi + \tan^{-1}\left|\frac{b}{a}\right| & if \ a < 0, b < 0 \\ |z| = \sqrt{a^2 + b^2}, \theta_z = -\tan^{-1}\left|\frac{b}{a}\right| & if \ a > 0, b < 0 \end{cases}$$

2.21. Based on the information given in the problem, we have:

$$a > 0 \tag{1}$$

$$Re\left(\frac{z-i}{z+i}\right) < 1 \tag{2}$$

$$Im\left(\frac{z-i}{z+i}\right) < a \tag{3}$$

The problem can be solved as follows:

$$\frac{z-i}{z+i} = \frac{x+iy-i}{x+iy+i} = \frac{x+i(y-1)}{x+i(y+1)}$$

$$= \frac{x+i(y-1)}{x+i(y+1)} \times \frac{x-i(y+1)}{x-i(y+1)} = \frac{x^2+y^2-1-2xi}{x^2+(y+1)^2}$$

$$= \frac{x^2+y^2-1}{x^2+(y+1)^2} - \frac{2x}{x^2+(y+1)^2}i \tag{4}$$

Solving (2) and (4):

$$\frac{x^2+y^2-1}{x^2+(y+1)^2} < 1 \Rightarrow x^2+y^2-1 < x^2+y^2+2y+1 \Rightarrow y > -1 \tag{5}$$

Equation (5) indicates the region which is above the line $y = -1$.

Solving (1), (3), and (4):

$$\frac{-2x}{x^2+(y+1)^2} < a \Rightarrow -\frac{2x}{a} < x^2+(y+1)^2 \Rightarrow x^2+(y+1)^2+\frac{2x}{a} > 0$$

$$\Rightarrow x^2+(y+1)^2+\frac{2x}{a}+\frac{1}{a^2} > \frac{1}{a^2}$$

$$\Rightarrow \left(x+\frac{1}{a}\right)^2+(y+1)^2 > \frac{1}{a^2} \tag{6}$$

Equation (6) indicates the outer part of a circle centered at $\left(-\frac{1}{a}, -1\right)$ with the radius equal to $\frac{1}{a}$.

From (5) and (6), it is noticed that the solution is the upper half and outer part of the circle $\left(x+\frac{1}{a}\right)^2+(y+1)^2 = \frac{1}{a^2}$. Choice (3) is the answer.

Notes

In this problem, the relation below has been used:

The equation of a circle centered at $(-a, -b)$ with the radius equal to c is as follows:

$$(x - a)^2 + (y - b)^2 = c^2$$

2.22. Based on the information given in the problem, we have:

$$e^z = -2 \tag{1}$$

$$\Rightarrow e^{x+iy} = e^x e^{iy} = e^x \cos y + i e^x \sin y = -2 \tag{2}$$

$$\Rightarrow \begin{cases} e^x \cos y = -2 \tag{3} \\ e^x \sin y = 0 \overset{e^x > 0}{\Rightarrow} \sin y = 0 \Rightarrow y = (2k+1)\pi, \quad k \epsilon \mathbb{Z} \end{cases} \tag{4}$$

Solving (3) and (4):

$$-e^x = -2 \Rightarrow e^x = 2 \Rightarrow x = \ln 2 \tag{5}$$

Solving (4), (5), and $z = x + iy$:

$$z = \ln 2 + i(2k+1)\pi, \quad k \epsilon \mathbb{Z}$$

As can be noticed, the equation has an infinite number of roots. Choice (4) is the answer.

Notes

In this problem, the relations below have been used:

$$e^{i\theta} = \cos \theta + i \sin \theta$$

$$\sin y = 0 \Rightarrow y = (2k+1)\pi, \quad k \epsilon \mathbb{Z}$$

$$\cos(2k+1)\pi = -1, \quad k \epsilon \mathbb{Z}$$

2.23. Based on the information given in the problem, we need to calculate the value of $z - \bar{z}$. Moreover, we have:

$$|z + ai| = |z + bi| \tag{1}$$

$$a \neq b, \quad a, b \epsilon \mathbb{R} \tag{2}$$

As we know:

$$z = x + iy \tag{3}$$

Therefore:

$$z - \bar{z} = (x + iy) - (x - iy) = 2yi \tag{4}$$

Solving (1) and (3):

$$|x + iy + ai| = |x + iy + bi| \tag{5}$$

$$\Rightarrow \sqrt{x^2 + (y+a)^2} = \sqrt{x^2 + (y+b)^2} \Rightarrow x^2 + (y+a)^2 = x^2 + (y+b)^2 \tag{6}$$

$$\Rightarrow (y+a)^2 = (y+b)^2 \Rightarrow y + a = \pm(y+b) \tag{7}$$

Solving (2) and (7):

$$y = -\left(\frac{a+b}{2}\right) \tag{8}$$

Solving (4) and (8):

$$z - \bar{z} = 2\left[-\left(\frac{a+b}{2}\right)\right]i \Rightarrow z - \bar{z} = -(a+b)i$$

Choice (1) is the answer.

Notes

In this problem, the relations below have been used:

$$z = (x + iy) \Rightarrow \bar{z} = (x - iy)$$

$$|a + ib| = \sqrt{a^2 + b^2}$$

2.24. Based on the information given in the problem, we have:

$$f(z) = z + \frac{1}{z} \tag{1}$$

$$Im\left\{z + \frac{1}{z}\right\} = 0 \tag{2}$$

As we know:

$$z = x + iy \tag{3}$$

Solving (1) and (3):

$$z + \frac{1}{z} = x + iy + \frac{1}{x + iy} = x + iy + \frac{1}{x + iy} \times \frac{x - iy}{x - iy}$$

$$\Rightarrow z + \frac{1}{z} = x + iy + \frac{x - iy}{x^2 + y^2} = x\left(1 + \frac{1}{x^2 + y^2}\right) + iy\left(1 - \frac{1}{x^2 + y^2}\right) \tag{4}$$

Solving (2) and (4):

$$1 - \frac{1}{x^2 + y^2} = 0 \Rightarrow x^2 + y^2 = 1$$

Thus, the region is a circle centered at the origin with the radius 1 unit. Choice (3) is the answer.

Notes

In this problem, the relation below has been used:

The equation of a circle centered at $(-a, -b)$ with the radius equal to c is as follows:

$$(x - a)^2 + (y - b)^2 = c^2$$

2.25. Based on the information given in the problem, we have:

$$z = \frac{2}{u + iv} \tag{1}$$

As we know:

$$z = x + iy \tag{2}$$

Solving (1) and (2):

$$x + iy = \frac{2}{u + iv} = \frac{2}{u + iv} \times \frac{u - iv}{u - iv}$$

$$\Rightarrow x + iy = 2\frac{u - iv}{u^2 + v^2} = \frac{2u}{u^2 + v^2} - i\frac{2v}{u^2 + v^2}$$

$$\Rightarrow x = \frac{2u}{u^2 + v^2}$$

Choice (3) is the answer.

2.26. Based on the information given in the problem, we need to calculate the value of $|z_1 - z_2|$ in which z_1 and z_2 are the roots of the quadratic equation $z^2 + z + 1 = i$.

As we know, the magnitude of the difference of roots in a quadratic equation with the from $az^2 + bz + c = 0$ can be calculated as follows:

$$|z_1 - z_2| = \left| \frac{\sqrt{b^2 - 4ac}}{a} \right|$$

Hence, for $z^2 + z + 1 - i = 0$, we have:

$$|z_1 - z_2| = \left| \sqrt{1 - 4(1-i)} \right| = \left| \sqrt{-3 + 4i} \right| = \left| \sqrt{5e^{i\theta}} \right| = \left| \sqrt{5}e^{i\frac{\theta}{2}} \right|$$

$$\Rightarrow |z_1 - z_2| = \sqrt{5}$$

Choice (3) is the answer.

Notes

In this problem, the relation below has been used:

If z_1 and z_2 are the roots of the quadratic equation $az^2 + bz + c = 0$, then:

$$|z_1 - z_2| = \left| \frac{\sqrt{b^2 - 4ac}}{a} \right|$$

2.27. Based on the information given in the problem, we have:

$$|z - 1| \leq Im(z)$$

The problem can be solved as follows:

$$|x + iy - 1| \leq y \Rightarrow |(x-1) + iy| \leq y$$

$$\Rightarrow (x-1)^2 + y^2 = y^2 \Rightarrow (x-1)^2 \leq 0$$

$$\Rightarrow x - 1 = 0 \Rightarrow x = 1$$

Choice (2) is the answer.

Notes

In this problem, the relations below have been used:

$$Im(z) = y$$

$$|a + ib| = \sqrt{a^2 + b^2}$$

2.28. Based on the information given in the problem, we have:

$$\frac{2z}{1+i} - \frac{2z}{i} = \frac{5}{2+i}$$

The problem can be solved as follows:

$$\frac{2iz - 2z(1+i)}{i(1+i)} = \frac{5}{2+i} \Rightarrow \frac{-2z}{i-1} = \frac{5}{2+i}$$

$$\Rightarrow -2z = \frac{5(i-1)}{2+i} \Rightarrow -2z = \frac{5(i-1)}{2+i} \times \frac{2-i}{2-i}$$

$$\Rightarrow -2z = \frac{5(i-1)(2-i)}{5} \Rightarrow -2z = 2i - i^2 - 2 + i$$

$$\Rightarrow -2z = 3i - 1 \Rightarrow z = \frac{1}{2}(1 - 3i)$$

Choice (1) is the answer.

Notes

In this problem, the relation below has been used:

$$i^2 = -1$$

2.29. Based on the information given in the problem, we have:

$$z\bar{z} = 36 \tag{1}$$

As we know:

$$z = x + iy \tag{2}$$

$$\bar{z} = x - iy \tag{3}$$

Solving (1, 2, and 3):

$$(x + iy)(x - iy) = 36 \Rightarrow x^2 - xiy + iyx - i^2y^2 = 36$$

$$\Rightarrow x^2 + y^2 = 36$$

The last relation denotes the equation of a circle with the radius 6 and the center at the origin. Choice (1) is the answer.

Notes

In this problem, the relations below have been used:

$$i^2 = -1$$

The equation of a circle centered at $(-a, -b)$ with the radius equal to c is as follows:

$$(x - a)^2 + (y - b)^2 = c^2$$

2.30. Based on the information given in the problem, we need to determine the equation of a circle with the radius four and the center located at $(-2, 1)$.

As we know, the equation of a circle with the radius R centered at z_0 is as follows:

$$|z - z_0| = R$$

Herein, $z_0 = x_0 + iy_0 = -2 + i$ and $R = 4$. Thus:

$$|z + 2 - i| = 4$$

Choice (1) is the answer.

2.31. Based on the information given in the problem, we have:

$$\left|\frac{z - 3i}{z + i}\right| = 1$$

As we know, $z = x + iy$. Thus:

$$\left|\frac{x + iy - 3i}{x + iy + i}\right| = 1$$

$$|x + iy - 3i| = |x + iy + i| \Rightarrow |x + i(y - 3)| = |x + i(y + 1)|$$

$$\Rightarrow x^2 + (y - 3)^2 = x^2 + (y + 1)^2$$

$$\Rightarrow x^2 + y^2 + 9 - 6y = x^2 + y^2 + 1 + 2y$$

$$\Rightarrow 8y = 8 \Rightarrow y = 1$$

Choice (4) is the answer.

Notes

In this problem, the relations below have been used:

$$\left|\frac{a}{b}\right| = \frac{|a|}{|b|}$$

$$|a + ib| = \sqrt{a^2 + b^2}$$

2.32. Based on the information given in the problem, we have:

$$|z - 2| = |z + 4|$$

As we know, $z = x + iy$. Hence:

$$|x + iy - 2| = |x + iy + 4| \Rightarrow |(x - 2) + iy| = |(x + 4) + iy|$$

$$\Rightarrow (x-2)^2 + y^2 = (x+4)^2 + y^2 \Rightarrow x^2 + 4 - 4x = x^2 + 16 + 8x$$

$$\Rightarrow x = -1$$

Choice (4) is the answer.

Notes

In this problem, the relation below has been used:

$$|a + ib| = \sqrt{a^2 + b^2}$$

2.33. Based on the information given in the problem, we have:

$$|z|^2 + 3\,Re(z^2) = 4$$

As we know, $z = x + iy$. Therefore:

$$\Rightarrow |x + iy|^2 + 3\,Re\left((x+iy)^2\right) = 4$$

$$\Rightarrow x^2 + y^2 + 3\,Re(x^2 + i^2y^2 + 2xiy) = 4$$

$$\Rightarrow x^2 + y^2 + 3\,Re(x^2 - y^2 + 2xiy) = 4$$

$$\Rightarrow x^2 + y^2 + 3(x^2 - y^2) = 4 \Rightarrow 4x^2 - 2y^2 = 4$$

$$\Rightarrow x^2 - \frac{y^2}{2} = 1$$

The solution shows the equation of hyperbola. Choice (1) is the answer.

Notes

In this problem, the relations below have been used:

$$i^2 = -1$$

$$|a + ib| = \sqrt{a^2 + b^2}$$

The general equation of hyperbola centered at $(-x_0, -y_0)$ is as follows:

$$\frac{(x - x_0)^2}{a^2} - \frac{(y - y_0)^2}{b^2} = 1$$

2.34. Based on the information given in the problem, we have:

$$\left|\frac{z+i}{z-i}\right| = \sqrt{2}$$

As we know, $z = x + iy$. Therefore:

$$\left|\frac{x+iy+i}{x+iy-i}\right| = \sqrt{2} \Rightarrow |x+i(y+1)| = \sqrt{2}|x+i(y-1)|$$

$$\Rightarrow x^2 + (y+1)^2 = 2\left(x^2 + (y-1)^2\right) \Rightarrow x^2 + y^2 + 2y + 1 = 2x^2 + 2y^2 - 4y + 2$$

$$\Rightarrow x^2 + y^2 - 6y + 1 = 0 \Rightarrow x^2 + y^2 - 6y + 9 = 8$$

$$\Rightarrow x^2 + (y-3)^2 = \left(2\sqrt{2}\right)^2$$

The solution shows the equation of a circle with the radius $2\sqrt{2}$ and the center at $(0, 3)$. Choice (3) is the answer.

Notes

In this problem, the relations below have been used:

$$\left|\frac{a}{b}\right| = \frac{|a|}{|b|}$$

$$|a+ib| = \sqrt{a^2 + b^2}$$

The equation of a circle with the radius R and the center at (a, b) is as follows:

$$(x-a)^2 + (y-b)^2 = R^2$$

2.35. Based on the information given in the problem, we have:

$$\left|\frac{z_1 - \bar{z}_2}{z_1 + \bar{z}_2}\right| = 1$$

$$\Rightarrow |z_1 - \bar{z}_2| = |z_1 + \bar{z}_2|$$

By assuming $z_1 = x_1 + iy_1$ and $z_2 = x_2 + iy_2$, we have:

$$\left|(x_1 + iy_1) - (x_2 + iy_2)\right| = \left|(x_1 + iy_1) + (x_2 + iy_2)\right|$$

$$\Rightarrow |(x_1 + iy_1) - x_2 + iy_2| = |(x_1 + iy_1) + x_2 - iy_2|$$

$$\Rightarrow |(x_1 - x_2) + i(y_1 + y_2)| = |(x_1 + x_2) + i(y_1 - y_2)|$$

$$\Rightarrow (x_1 - x_2)^2 + (y_1 + y_2)^2 = (x_1 + x_2)^2 + (y_1 - y_2)^2$$

$$\Rightarrow x_1^2 + x_2^2 - 2x_1x_2 + y_1^2 + y_2^2 + 2y_1y_2 = x_1^2 + x_2^2 + 2x_1x_2 + y_1^2 + y_2^2 - 2y_1y_2$$

$$\Rightarrow x_1x_2 - y_1y_2 = 0 \Rightarrow Re(z_1z_2) = 0$$

Choice (3) is the answer.

Notes

In this problem, the relations below have been used:

$$\left| \frac{a}{b} \right| = \frac{|a|}{|b|}$$

$$\overline{a + ib} = a - ib$$

$$Re(z_1z_2) = x_1x_2 - y_1y_2$$

2.36. Based on the information given in the problem, we have:

$$u(x, y) + iv(x, y) = \frac{1}{1 - z}$$

As we know, $z = x + iy$. Therefore:

$$u(x, y) + iv(x, y) = \frac{1}{1 - (x + iy)}$$

$$\Rightarrow u(x, y) + iv(x, y) = \frac{1}{(1 - x) - iy} \times \frac{(1 - x) + iy}{(1 - x) + iy}$$

$$\Rightarrow u(x, y) + iv(x, y) = \frac{(1 - x) + iy}{(1 - x)^2 + y^2}$$

$$\Rightarrow u(x, y) + iv(x, y) = \frac{1 - x}{(1 - x)^2 + y^2} + i\frac{y}{(1 - x)^2 + y^2}$$

$$\Rightarrow u(x, y) = \frac{1 - x}{(1 - x)^2 + y^2}, \quad v(x, y) = \frac{y}{(1 - x)^2 + y^2}$$

Choice (4) is the answer.

Notes

In this problem, the relation below has been used:

$$i^2 = -1$$

2.4   Holomorphic (Analytic) Function and Its Harmonic Conjugate Functions

2.37. A function $u(x, y)$ is said to have a harmonic conjugate function $v(x, y)$ if and only if they are, respectively, the real and imaginary parts of a holomorphic function $f(z)$ of the complex variable $z = x + iy$. In other words, $v(x, y)$ is the harmonic conjugate function of $u(x, y)$ if $f(z) = u(x, y) + iv(x, y)$, and $u(x, y)$ and $v(x, y)$ satisfy the Cauchy-Riemann equations, that is:

$$u_x(x, y) = v_y(x, y) \tag{1}$$

$$u_y(x, y) = -v_x(x, y) \tag{2}$$

Based on the information given in the problem, we have:

$$u(x, y) = 2x(3 - y) \tag{3}$$

The harmonic conjugate function of the function can be calculated as follows:

Solving (1) and (3):

$$\frac{\partial}{\partial x}(2x(3 - y)) = v_y(x, y) \tag{4}$$

$$\Rightarrow v_y(x, y) = 2(3 - y) \tag{5}$$

$$\Rightarrow v(x, y) = \int 2(3 - y)dy = 6y - y^2 + h(x) \tag{6}$$

On the other hand, by solving (2), (3), and (6), we have:

$$-2x = -h(x) \tag{7}$$

$$\Rightarrow h(x) = x^2 + c_1 \tag{8}$$

Solving (6) and (8):

$$v(x, y) = 6y - y^2 + x^2 + c_1 \tag{9}$$

$$\Rightarrow v(x, y) = 6y - y^2 - 9 + 9 + x^2 + c_1 \tag{10}$$

$$\Rightarrow v(x, y) = -(3 - y)^2 + x^2 + c_1 + 9 \tag{11}$$

If we assume that $c_1 + 9 = c$, then:

$$v(x, y) = x^2 - (3 - y)^2 + c$$

Choice (4) is the answer.

2.38. Based on the information given in the problem, we have:

$$u(x, y) = \ln(x^2 + y^2) \tag{1}$$

Herein, $u(x, y)$ and its harmonic conjugate function, that is, $v(x, y)$, must satisfy the Cauchy-Riemann equations. In other words:

$$u_x(x, y) = v_y(x, y) \tag{2}$$

$$u_y(x, y) = -v_x(x, y) \tag{3}$$

Solving (1) and (2):

$$\frac{\partial}{\partial x}\left(\ln\left(x^2 + y^2\right)\right) = v_y(x, y) \tag{4}$$

$$\Rightarrow v_y(x, y) = \frac{2x}{x^2 + y^2} \tag{5}$$

$$\Rightarrow v(x, y) = \int \frac{2x}{x^2 + y^2}\, dy = 2x\frac{1}{x}\arctan\frac{y}{x} + h(x) = 2\arctan\frac{y}{x} + h(x) \tag{6}$$

On the other hand, by solving (1), (3), and (6), we have:

$$\frac{2y}{x^2 + y^2} = -\left(2\frac{-\frac{y}{x^2}}{1 + \frac{y^2}{x^2}} + h(x)\right) \tag{7}$$

$$\Rightarrow \frac{2y}{x^2 + y^2} = \frac{2y}{x^2 + y^2} - h(x) \Rightarrow h(x) = 0 \Rightarrow h(x) = c \tag{8}$$

Solving (6) and (8):

$$v(x, y) = 2\arctan\frac{y}{x} + c = 2\tan^{-1}\frac{y}{x} + c$$

Choice (3) is the answer.

Notes

In this problem, the relations below have been used:

$$\frac{d}{dx}\ln(u(x)) = \frac{u(x)}{u(x)}$$

$$\int \frac{dx}{a^2 + x^2} = \frac{1}{a}\arctan\frac{x}{a} + c$$

2.39. Based on the information given in the problem, we need to determine the values of a and b so that the function below is a harmonic function.

$$u(x, y) = x^2 + ay^2 + bxy \tag{1}$$

As we know, a function $u(x, y) = x^2 + ay^2 + bxy$ is a harmonic function if the following relation is held about that:

$$u_{xx} + u_{yy} = 0 \tag{2}$$

Solving (1) and (2):

$$\begin{cases} u_x = 2x + by \Rightarrow u_{xx} = 2 \\ u_y = 2ay + bx \Rightarrow u_{yy} = 2a \end{cases} \qquad \begin{matrix} (3) \\ (4) \end{matrix}$$

Solving (2, 3, and 4):

$$2 + 2a = 0 \Rightarrow a = -1$$

The value of b is arbitrary. In other words:

$$b \in \mathbb{R}$$

Choice (2) is the answer.

Notes

In this problem, the relation below has been used:

A function $u(x, y)$ is a harmonic function if $u_{xx} + u_{yy} = 0$.

2.40. Based on the information given in the problem, we know that $u(x, y)$ is a harmonic function. Thus:

$$\frac{\partial^2 u(x, y)}{\partial x^2} + \frac{\partial^2 u(x, y)}{\partial y^2} = 0 \qquad (1)$$

Moreover, we have:

$$\frac{\partial^2 u(x, y)}{\partial z \partial \bar{z}} \qquad (2)$$

As we know, $z = x + iy$ and $\bar{z} = x - iy$. Therefore:

$$x = \frac{z + \bar{z}}{2} \Rightarrow \frac{\partial x}{\partial z} = \frac{1}{2}, \quad \frac{\partial x}{\partial \bar{z}} = \frac{1}{2} \qquad (3)$$

$$y = \frac{z - \bar{z}}{2i} \Rightarrow \frac{\partial y}{\partial z} = \frac{1}{2i}, \quad \frac{\partial y}{\partial \bar{z}} = -\frac{1}{2i} \qquad (4)$$

By applying the chain rule for the first-order partial derivative, we have:

$$\frac{\partial u(x, y)}{\partial \bar{z}} = \frac{\partial u(x, y)}{\partial x} \frac{\partial x}{\partial \bar{z}} + \frac{\partial u(x, y)}{\partial y} \frac{\partial y}{\partial \bar{z}} \qquad (5)$$

Solving (3, 4, and 5):

$$\frac{\partial u(x, y)}{\partial \bar{z}} = \frac{1}{2} \frac{\partial u(x, y)}{\partial x} - \frac{1}{2i} \frac{\partial u(x, y)}{\partial y} \qquad (6)$$

By applying the chain rule for the second-order partial derivative, we have:

$$\frac{\partial^2 u(x, y)}{\partial z \partial \bar{z}} = \frac{\partial}{\partial z}\left(\frac{\partial u(x, y)}{\partial \bar{z}}\right) \tag{7}$$

$$\Rightarrow \frac{\partial^2 u(x, y)}{\partial z \partial \bar{z}} = \frac{\partial}{\partial x}\left(\frac{\partial u(x, y)}{\partial \bar{z}}\right)\frac{\partial x}{\partial z} + \frac{\partial}{\partial y}\left(\frac{\partial u(x, y)}{\partial \bar{z}}\right)\frac{\partial y}{\partial z} \tag{8}$$

Solving (3), (4), (6), and (8):

$$\frac{\partial^2 u(x, y)}{\partial z \partial \bar{z}} = \frac{\partial}{\partial x}\left(\frac{1}{2}\frac{\partial u(x, y)}{\partial x} - \frac{1}{2i}\frac{\partial u(x, y)}{\partial y}\right)\frac{1}{2} + \frac{\partial}{\partial y}\left(\frac{1}{2}\frac{\partial u(x, y)}{\partial x} - \frac{1}{2i}\frac{\partial u(x, y)}{\partial y}\right)\frac{1}{2i} \tag{9}$$

$$\Rightarrow \frac{\partial^2 u(x, y)}{\partial z \partial \bar{z}} = \frac{1}{4}\frac{\partial^2 u(x, y)}{\partial x^2} - \frac{1}{4i}\frac{\partial^2 u(x, y)}{\partial x \partial y} + \frac{1}{4i}\frac{\partial^2 u(x, y)}{\partial y \partial x} - \frac{\partial^2 u(x, y)}{4i^2 \partial y^2} \tag{10}$$

$$\Rightarrow \frac{\partial^2 u(x, y)}{\partial z \partial \bar{z}} = \frac{1}{4}\left(\frac{\partial^2 u}{\partial x^2} + \frac{\partial^2 u}{\partial y^2}\right) \tag{11}$$

Solving (1) and (11):

$$\Rightarrow \frac{\partial^2 u(x, y)}{\partial z \partial \bar{z}} = 0$$

Choice (1) is the answer.

Notes

In this problem, the relations below have been used:

$$\frac{\partial u(x, y)}{\partial z} = \frac{\partial u(x, y)}{\partial x}\frac{\partial x}{\partial z} + \frac{\partial u(x, y)}{\partial y}\frac{\partial y}{\partial z}$$

$$i^2 = -1$$

2.41. Based on the information given in the problem, we have:

$$u(x, y) = x^3 + \alpha x^2 y + \beta x y^2 + y^3 \tag{1}$$

We know that $u(x, y)$ is a harmonic function. Hence:

$$u_{xx}(x, y) + u_{yy}(x, y) = 0 \tag{2}$$

The problem can be solved as follows:

$$u_x(x, y) = 3x^2 + 2\alpha x y + \beta y^2 \Rightarrow u_{xx} = 6x + 2\alpha y \tag{3}$$

$$u_y(x, y) = \alpha x^2 + 2\beta x y + 3y^2 \Rightarrow u_{yy} = 2\beta x + 6y \tag{4}$$

Solving (2, 3, and 4):

$$6x + 2\alpha y + 6y + 2\beta x = 0 \tag{5}$$

$$\Rightarrow (2\beta + 6)x + (2\alpha + 6)y = 0 \tag{6}$$

$$\Rightarrow \alpha = \beta = -3$$

Choice (1) is the answer.

2.42. Based on the information given in the problem, we have:

$$u(x, y) = x + e^x \cos y \tag{1}$$

As we know:

$$f'(z) = u_x(x, y) - iu_y(x, y) \tag{2}$$

Solving (1) and (2):

$$f'(z) = 1 + e^x \cos y + ie^x \sin y \tag{3}$$

$$\Rightarrow f'(1) = f'(1 + 0i) = 1 + e^1 \cos 0 + ie^1 \sin 0 \tag{4}$$

$$\Rightarrow f'(1) = 1 + e$$

Choice (1) is the answer.

Notes

In this problem, the relations below have been used:

$$\frac{d}{dx} e^x = e^x$$

$$\frac{d}{dx} \cos y = -\sin y$$

$$\cos 0 = 1$$

$$\sin 0 = 0$$

References

1. Rahmani-Andebili, M. (2024). Precalculus (2nd Ed.) – Practice Problems, Methods, and Solutions, Springer Nature.
2. Rahmani-Andebili, M. (2023). Calculus III – Practice Problems, Methods, and Solutions, Springer Nature.
3. Rahmani-Andebili, M. (2023). Calculus II – Practice Problems, Methods, and Solutions, Springer Nature.
4. Rahmani-Andebili, M. (2023). Calculus I (2nd Ed.) – Practice Problems, Methods, and Solutions, Springer Nature.
5. Rahmani-Andebili, M. (2022). Differential Equations – Practice Problems, Methods, and Solutions, Springer Nature.
6. Rahmani-Andebili, M. (2021). Calculus – Practice Problems, Methods, and Solutions, Springer Nature.
7. Rahmani-Andebili, M. (2021). Precalculus – Practice Problems, Methods, and Solutions, Springer Nature.

Abstract

In this chapter, the basic and advanced problems of complex transformations are presented. The subjects include linear, power, reciprocal, exponential, natural logarithm, hyperbolic sine and cosine, sine and cosine, and linear fractional complex transformation. Herein, different types of problems and exercises are presented that are categorized as follows:

○ *Problems with detailed solution*: They have been designed to teach students the subjects in detail. Moreover, they have been categorized into different levels based on their difficulty levels (easy, normal, and hard) and calculation amounts (small, normal, and large).

○ *Partially solved exercises*: They have been designed to encourage students to practice more problems while guiding them through the problem-solving procedure and hinting the required formulas.

○ *Exercises with final answer*: They have been designed to encourage students to practice by themselves while hinting them by the final answer as well as to help instructors to give tests or quizzes.

3.1   Linear Complex Transformation

3.1. Find the image of the upper half plane in z-plane under the transformation $w = (1 + i)z$ [1–7].

Difficulty level   ○ Easy   ● Normal   ○ Hard
Calculation amount   ● Small   ○ Normal   ○ Large

1) $u + v = 1$
2) $u < v$
3) $u + v = -1$
4) $u > v$

Partially Solved Exercise

Find the image of the right half plane in z-plane under the transformation $w = i(z + 1)$.

Solution

Based on the information given in the problem, we have:

$$x > 0 \tag{1}$$

$$w = i(z + 1) \tag{2}$$

As we know, $z = x + iy$ and $w = u + iv$. Therefore:

$$w = i(\qquad) = (\qquad) + i(\qquad)$$

$$\Rightarrow \begin{cases} u = (\qquad) & (3) \\ v = (\qquad) & (4) \end{cases}$$

Solving (1) and (4):

$$v > 1$$

Exercise

Find the image of x-axis in z-plane under the transformation $w = (1 + i)z$.

Final Answer

$u = v$

Exercise

Find the image of the upper half plane in z-plane under the transformation $w = i(z + 1)$.

Final Answer

The left half plane in w-plane

3.2. Determine the transformation that maps the region shown in Fig. 3.1 to the upper half plane in w-plane ($v > 0$).

Difficulty level   ○ Easy   ● Normal   ○ Hard
Calculation amount   ○ Small   ● Normal   ○ Large

1) $w = [z - (1 + i)]^3$
2) $w = [z - (1 + i)]^6$
3) $w = -[z - (1 + i)]^3$
4) $w = -[z - (1 + i)]^6$

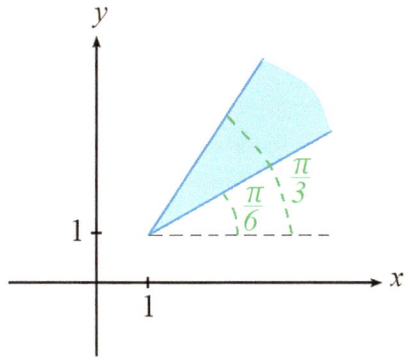

Fig. 3.1 The region in z-plane related to Problem 3.2

Exercise

Determine the transformation that maps the region shown in Fig. 3.1 to the whole plane in w-plane.

Final Answer

$w = [z - (1 + i)]^{12}$

3.2   Power Complex Transformation

3.3. Find the image of the region described by $\{x > 0,\ \ y > 0,\ \ xy > 1\}$ in z-plane under the transformation $w = z^2$.
   Difficulty level   ○ Easy   ● Normal   ○ Hard
   Calculation amount   ● Small   ○ Normal   ○ Large
   1) $v > 1$
   2) $v > 2$
   3) $u > 1$
   4) $uv > 1$

3.4. Find the image of the circle $x^2 + y^2 = a^2$ in z-plane under the transformation $w = z^2$.
   Difficulty level   ● Easy   ○ Normal   ○ Hard
   Calculation amount   ● Small   ○ Normal   ○ Large
   1) $u^2 + v^2 = a^2$
   2) $u^2 + v^2 = a^4$
   3) $u^4 + v^4 = a^2$
   4) $u^4 + v^4 = a^4$

Exercise

Find the image of the circle $x^2 + y^2 = 1$ in z-plane under the transformation $w = z^3$.

Final Answer

A unit circle centered at the origin in w-plane

3.5. Determine the image of the line $Re(z) = 1$ in z-plane under the transformation $w = z^2$.
   Difficulty level   ○ Easy   ● Normal   ○ Hard
   Calculation amount   ● Small   ○ Normal   ○ Large
   1) $u = 2y,\quad v = 1 + y^2$
   2) $u = 2y,\quad v = 1 - y^2$
   3) $u = 1 + y^2,\quad v = 2y$
   4) $u = 1 - y^2,\quad v = 2y$

3.6. Which one of the following regions is the image of the region described by $\{0 \le \theta \le \pi\}$ in z-plane under the transformation $w = z^{\frac{1}{3}}$?
   Difficulty level   ● Easy   ○ Normal   ○ Hard
   Calculation amount   ● Small   ○ Normal   ○ Large

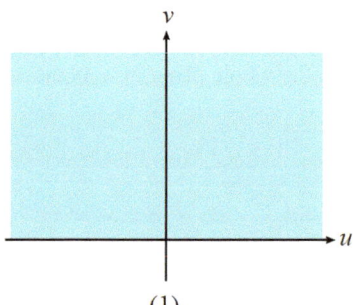

(1)

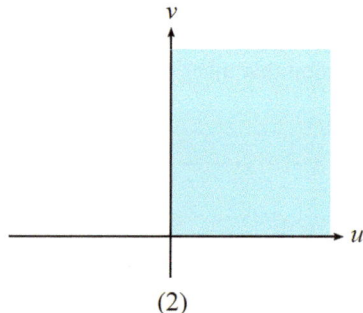

(2)

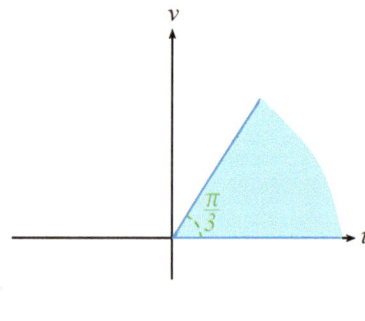

(3)

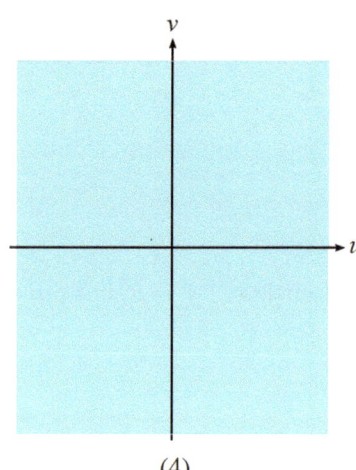

(4)

Exercise

Determine the image of the region described by $\{0 \le \theta \le \pi\}$ in z-plane under the transformation $w = \sqrt{z}$.

Final Answer
The first quadrant in w-plane

3.7. Which one of the following regions is the image of the region (shown in Fig. 3.2) under the transformation $w = z^4$?

Difficulty level   ● Easy   ○ Normal   ○ Hard
Calculation amount   ● Small   ○ Normal   ○ Large

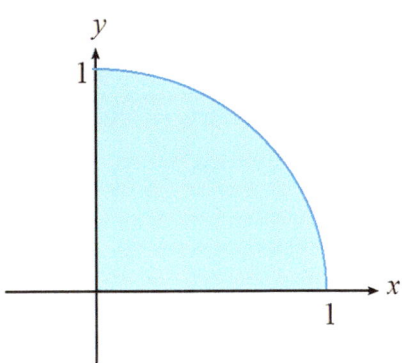

Fig. 3.2 The region in z-plane related to Problem 3.7

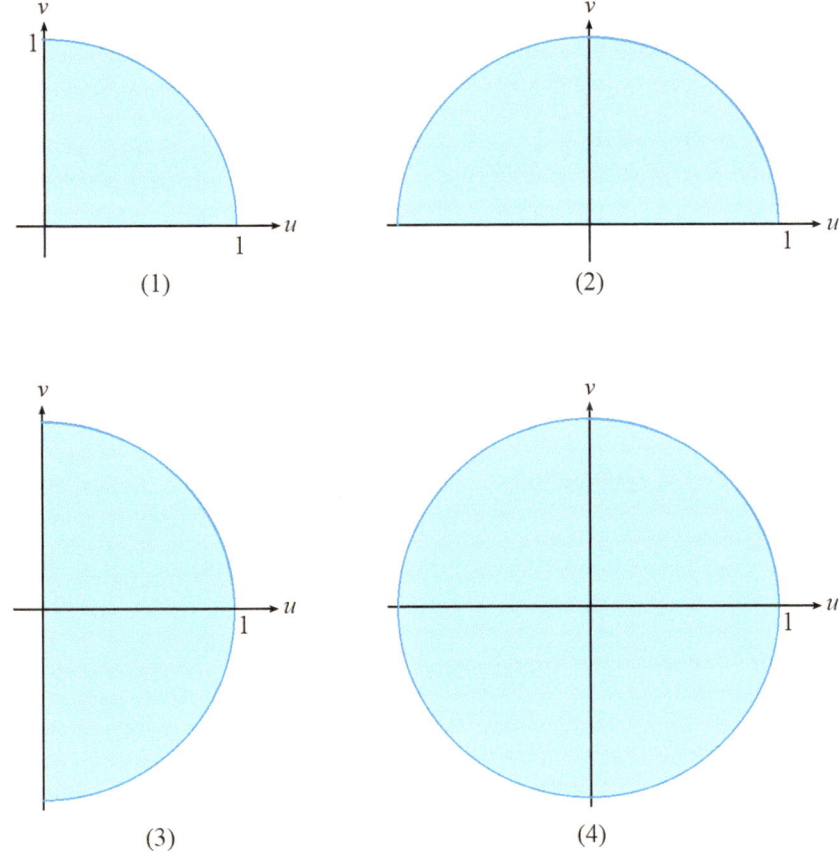

Exercise

Determine the image of the first quadrant in z-plane under the transformation $w = z^2$.

Final Answer
The upper half plane in w-plane

3.3   Reciprocal Complex Transformation

3.8. What is the image of the point (x, y) in z-plane under the transformation $w = \frac{1}{z}$?

Difficulty level   ● Easy   ○ Normal   ○ Hard

Calculation amount   ○ Small   ● Normal   ○ Large

1) $u = \dfrac{x}{x^2 + y^2}, \quad v = \dfrac{-y}{x^2 + y^2}$

2) $u = \dfrac{x}{x^2 + y^2}, \quad v = \dfrac{y}{x^2 + y^2}$

3) $u = \dfrac{y}{x^2 + y^2}, \quad v = \dfrac{x}{x^2 + y^2}$

4) $u = \dfrac{y}{x^2 + y^2}, \quad v = \dfrac{-x}{x^2 + y^2}$

3.9. Determine the image of the line $x = 1$ in z-plane under the transformation $w = \frac{1}{z}$.

Difficulty level   ○ Easy   ● Normal   ○ Hard

Calculation amount   ○ Small   ● Normal   ○ Large

1) A circle with the radius $\frac{1}{2}$ and the center at $\left(\frac{1}{2}, 0\right)$

2) A circle with the radius $\frac{1}{2}$ and the center at $\left(-\frac{1}{2}, 0\right)$

3) A circle with the radius $\frac{1}{2}$ and the center at $\left(0, \frac{1}{2}\right)$

4) A circle with the radius 1 and the center at $\left(\frac{1}{2}, 0\right)$

Partially Solved Exercise

Determine the image of the line $y = 1$ in z-plane under the transformation $w = \frac{1}{z}$.

Solution

Based on the information given in the problem, the region in z-plane under transformation and the transformation are as follows:

$$y = 1 \tag{1}$$

$$w = \frac{1}{z} \Rightarrow z = \frac{1}{w} \tag{2}$$

As we know, $z = x + iy$ and $w = u + iv$. Therefore:

$$x + iy = \frac{1}{u + iv} \tag{3}$$

$$\Rightarrow x + iy = \left(\frac{\rule{1cm}{0.4pt}}{}\right) \times \left(\frac{\rule{1cm}{0.4pt}}{}\right) = \left(\frac{\rule{1cm}{0.4pt}}{}\right) \tag{4}$$

$$\Rightarrow x + iy = \left(\frac{\rule{1cm}{0.4pt}}{}\right) + i\left(\frac{\rule{1cm}{0.4pt}}{}\right)$$

$$\Rightarrow \begin{cases} x = \left(\underline{\hspace{2cm}} \right) \\ y = \left(\underline{\hspace{2cm}} \right) \end{cases}$$

(5)

(6)

Solving (1) and (6):

$$1 = \left(\underline{\hspace{2cm}} \right) \Rightarrow \left(\underline{\hspace{3cm}} \right) = 0$$

(7)

$$\Rightarrow \left(\underline{\hspace{3cm}} \right) + \frac{1}{4} - \frac{1}{4} = 0$$

$$\Rightarrow u^2 + \left(v + \frac{1}{2} \right)^2 = \frac{1}{4}$$

(8)

The relation (8) shows the equation of a circle with the radius $\frac{1}{2}$ and the center at $\left(0, -\frac{1}{2}\right)$.

Notes

In this problem, the relation below has been used.

The equation of a circle with the radius R and the center at (x_0, y_0) is as follows.

$$(x - x_0)^2 + (y - y_0)^2 = R^2$$

3.10. Find the image of the circle $|z - 1| = 1$ in z-plane under the transformation $w = \frac{i}{z}$.

Difficulty level   ○ Easy   ● Normal   ○ Hard

Calculation amount   ○ Small   ● Normal   ○ Large

1) A line parallel to the imaginary axis in w-plane
2) A line parallel to the real axis in w-plane
3) A circle with the radius $\frac{1}{2}$ and the center at $\left(0, -\frac{i}{2}\right)$ in w-plane
4) A circle with the radius $\frac{1}{2}$ and the center at $\left(0, \frac{i}{2}\right)$ in w-plane

3.11. Which one of the following regions is the image of the region (illustrated in Fig. 3.3) under the transformation $w = \frac{1}{z}$?

Difficulty level   ○ Easy   ○ Normal   ● Hard

Calculation amount   ○ Small   ● Normal   ○ Large

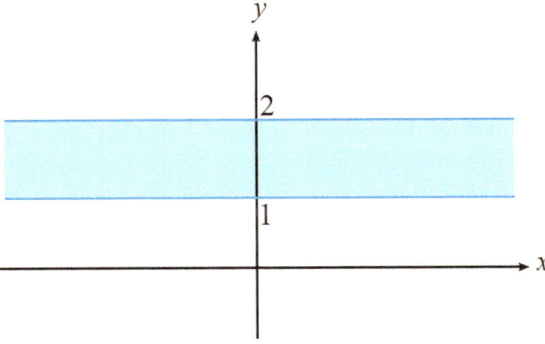

Fig. 3.3 The region in z-plane related to Problem 3.11

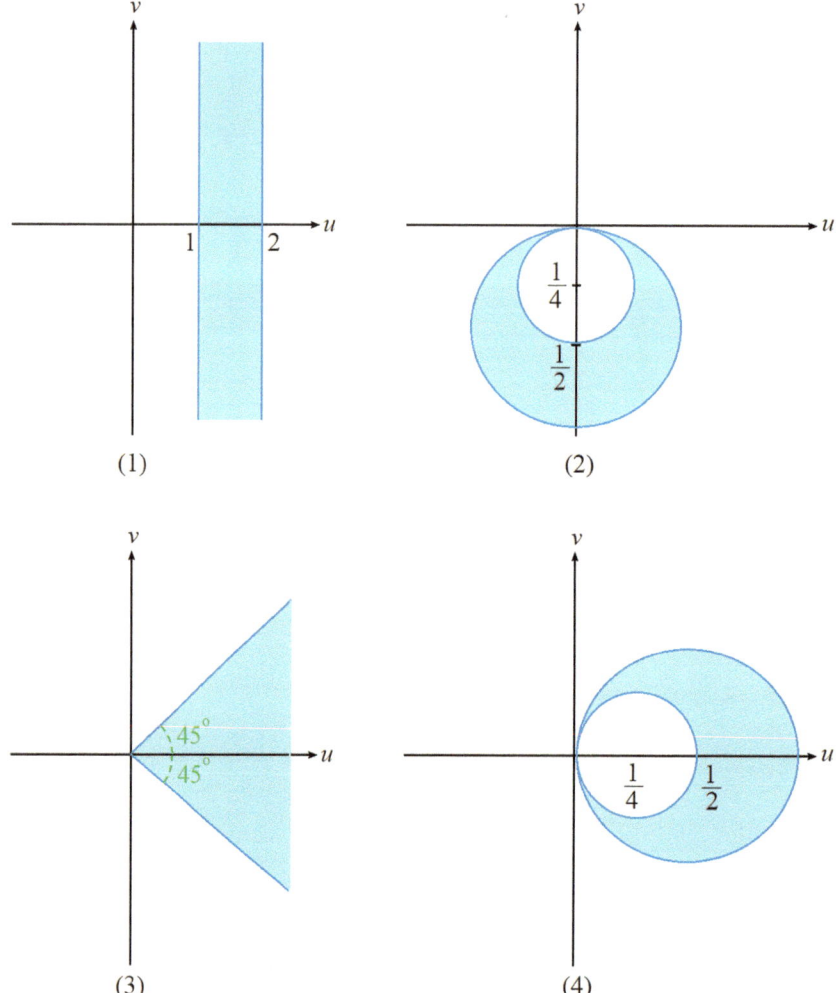

(1)   (2)

(3)   (4)

3.12. Find the image of the circle $(x + 1)^2 + (y - 1)^2 = 2$ in z-plane under the transformation $w = \frac{1}{z}$.

Difficulty level   ○ Easy   ● Normal   ○ Hard
Calculation amount   ○ Small   ○ Normal   ● Large

1) $v = u + \frac{1}{2}$
2) $u + v + \frac{1}{2} = 0$
3) $(u + 1)^2 + (v - 1)^2 = \frac{1}{2}$
4) $(u - 1)^2 + (v + 1)^2 = \frac{1}{2}$

3.13. What is the image of the line $2y + 3x - 1 = 0$ in z-plane under the transformation $w = \frac{1}{z}$?

Difficulty level   ○ Easy   ○ Normal   ● Hard
Calculation amount   ○ Small   ● Normal   ○ Large

1) A circle in which the origin is outside of that
2) A circle that passes from the origin
3) A circle in which the origin is inside of that
4) None of the abovementioned choices

3.14. What is the image of the unit circle $z = e^{i\theta}$, $0 \le \theta \le 2\pi$ in z-plane under the transformation $w = z + \frac{1}{z}$.

Difficulty level ○ Easy ○ Normal ● Hard
Calculation amount ○ Small ● Normal ○ Large

1) A circle
2) An ellipse
3) The line segment $-1 \le v \le 1$
4) The line segment $-2 \le u \le 2$

3.4 Exponential Complex Transformation

3.15. Which one of the following regions is similar to the image of the region described by $\left\{1 \le x \le 2, \ \frac{\pi}{4} \le y \le \frac{\pi}{2}\right\}$ in z-plane under the transformation $w = e^z$?

Difficulty level ○ Easy ● Normal ○ Hard
Calculation amount ● Small ○ Normal ○ Large

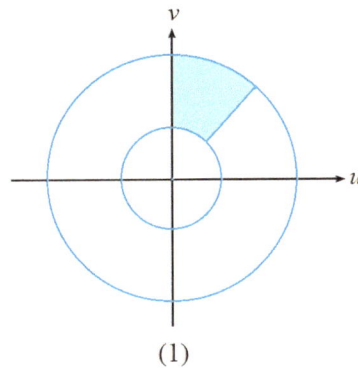

(1)

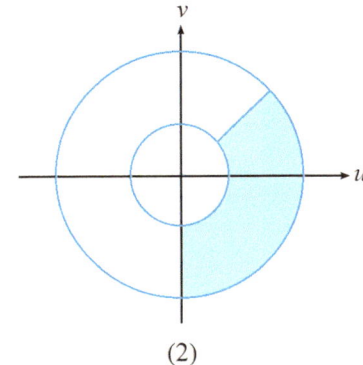

(2)

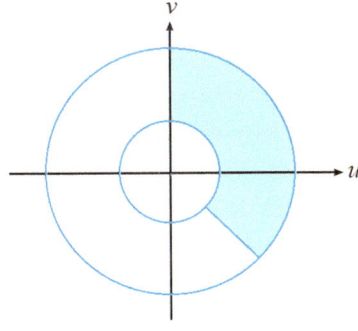

(3)

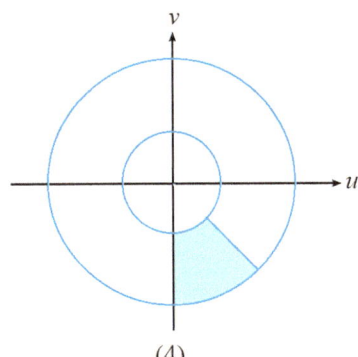

(4)

3.16. Which one of the following regions is the image of the region described by $\{x \le 0, \ 0 \le y \le \pi\}$ under the transformation $w = e^z$?

Difficulty level ○ Easy ● Normal ○ Hard
Calculation amount ● Small ○ Normal ○ Large

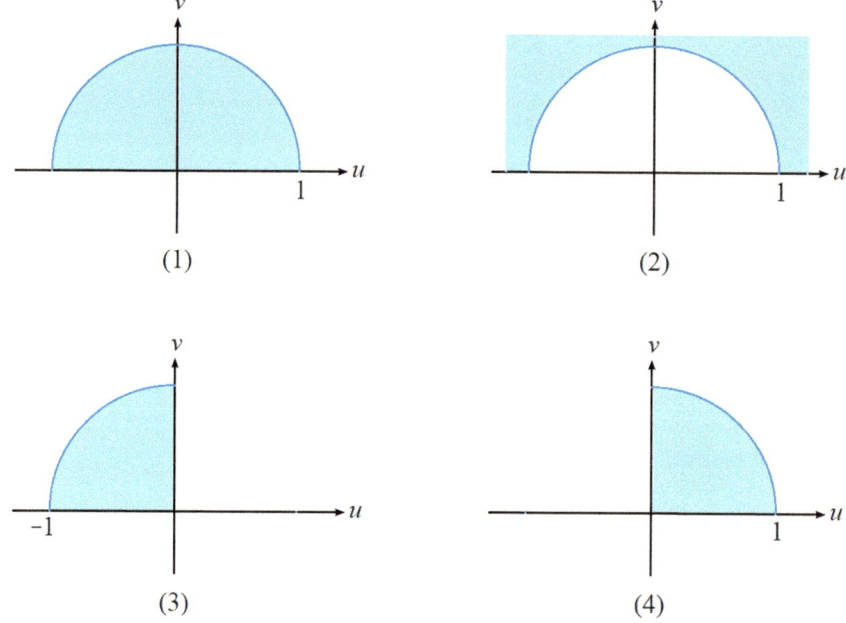

Exercise

Determine the image of the region described by $\{x \geq 0, \ 0 \leq y \leq 2\pi\}$ under the transformation $w = e^z$.

Final Answer

The outer region of the unit circle centered at the origin

3.17. Which one of the following regions is the image of the region described by $\left\{ Re\{z\} \leq 0, \ 0 \leq Im\{z\} \leq \frac{\pi}{2} \right\}$ in z-plane under the transformation $w = e^z$?

Difficulty level   ○ Easy   ● Normal   ○ Hard
Calculation amount   ● Small   ○ Normal   ○ Large

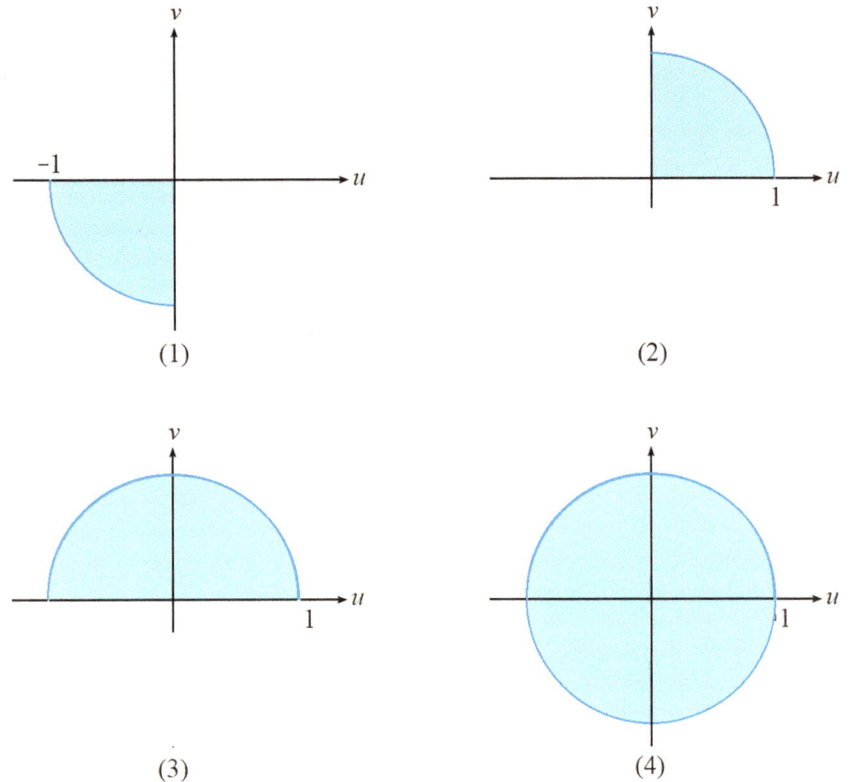

(1)   (2)

(3)   (4)

Determine the image of the region described by $\{Re\{z\} \leq 0, \ \ 0 \leq Im\{y\} \leq 2\pi\}$ under the transformation $w = e^z$.

Final Answer
The inner region of the unit circle centered at the origin

3.5   Natural Logarithm Complex Transformation

3.18. Which one of the following regions is the image of the region described by $\{x > 1, \ \ y \geq 0\}$ in z-plane under the transformation $w = \ln(z - 1)$ which is the principal branch of the natural logarithm?
    Difficulty level   ○ Easy   ○ Normal   ● Hard
    Calculation amount   ○ Small   ● Normal   ○ Large

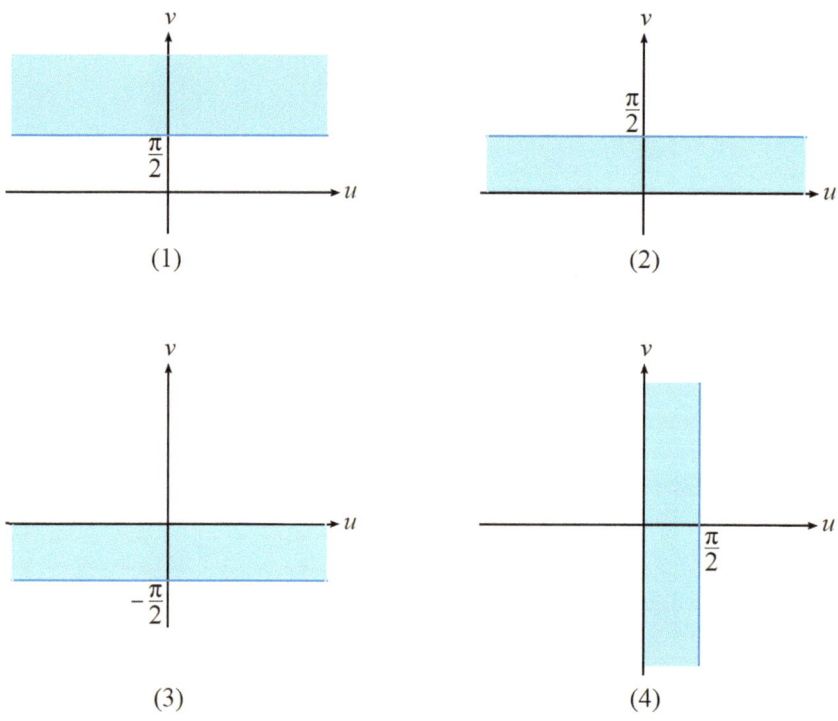

3.6 Hyperbolic Sine and Cosine Complex Transformation

3.19. Find the image of the line $y = \frac{\pi}{4}$ in z-plane under the transformation $w = \cosh z$.

    Difficulty level   ○ Easy   ● Normal   ○ Hard

    Calculation amount   ○ Small   ● Normal   ○ Large

    1) $u^2 - v^2 = 1$

    2) $v^2 - u^2 = 1$

    3) $u^2 - v^2 = \frac{1}{2}$

    4) $v^2 - u^2 = \frac{1}{2}$

3.7 Sine and Cosine Complex Transformation

3.20. Find the image of the line $x = \frac{\pi}{6}$ in z-plane under the transformation $w = \sin z$.

    Difficulty level   ○ Easy   ● Normal   ○ Hard

    Calculation amount   ○ Small   ● Normal   ○ Large

    1) A line

    2) An ellipse

    3) A circle

    4) A hyperbola

3.8　Linear Fractional Complex Transformation

3.21. Find a linear fractional transformation that maps the three points $z_1 = 0$, $z_2 = -i$, and $z_3 = -1$ in z-plane to $w_1 = i$, $w_2 = 1$, and $w_3 = 0$ in w-plane.

Difficulty level　○ Easy　● Normal　○ Hard
Calculation amount　○ Small　● Normal　○ Large

1) $w = i\left(\frac{z+1}{z-1}\right)$

2) $w = -i\left(\frac{z+1}{z-1}\right)$

3) $w = i\left(\frac{z-1}{z+1}\right)$

4) $w = -i\left(\frac{z-1}{z+1}\right)$

References

1. Rahmani-Andebili, M. (2024). Precalculus (2nd Ed.) – Practice Problems, Methods, and Solutions, Springer Nature.
2. Rahmani-Andebili, M. (2023). Calculus III – Practice Problems, Methods, and Solutions, Springer Nature.
3. Rahmani-Andebili, M. (2023). Calculus II – Practice Problems, Methods, and Solutions, Springer Nature.
4. Rahmani-Andebili, M. (2023). Calculus I (2nd Ed.) – Practice Problems, Methods, and Solutions, Springer Nature.
5. Rahmani-Andebili, M. (2022). Differential Equations – Practice Problems, Methods, and Solutions, Springer Nature.
6. Rahmani-Andebili, M. (2021). Calculus – Practice Problems, Methods, and Solutions, Springer Nature.
7. Rahmani-Andebili, M. (2021). Precalculus – Practice Problems, Methods, and Solutions, Springer Nature.

Abstract

In this chapter, the problems of the third chapter are fully solved, in detail, step-by-step, and with different methods.

4.1   Linear Complex Transformation

4.1. Based on the information given in the problem, the transformation and the region in z-plane under transformation are as follows [1–7]:

$$w = (1+i)z \tag{1}$$

$$y > 0 \tag{2}$$

As we know, $z = x + iy$ and $w = u + iv$. Hence:

$$w = (1+i)(x+iy) = x - y + i(x+y) \Rightarrow \begin{cases} u = x - y \\ v = x + y \end{cases} \tag{3}$$

Since $y > 0$, we can conclude that:

$$x - y < x < x + y \tag{4}$$

Solving (3) and (4):

$$u < x < v \tag{5}$$

$$\Rightarrow u < v$$

Choice (2) is the answer.

4.2. Based on the information given in the problem, the transformed region is $v > 0$. Moreover, the region in z-plane is shown in Fig. 4.1.

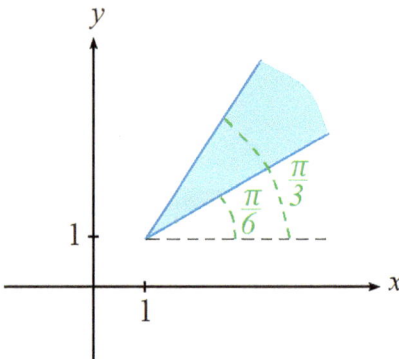

Fig. 4.1 The region in z-plane before the complex transformation

As we know, the region $0 \leq \theta < \frac{\pi}{n}$ in polar coordinates under the transformation $w = z^n$ is mapped on the upper half plane in w-plane.

Thus, the region in z-plane (illustrated in Fig. 4.1) needs to be moved to the origin first. This transformation can be done as follows:

$$w_1 = z - (1 + i)$$

The result is shown in Fig. 4.2.

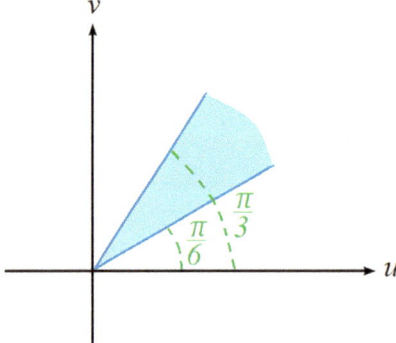

Fig. 4.2 The region in w-plane after the first complex transformation

Then, the transformed region must be rotated clockwise about $\frac{\pi}{6}$ radians. This transformation can be performed as follows:

$$w_2 = e^{-i\frac{\pi}{6}} w_1$$

The result is shown in Fig. 4.3.

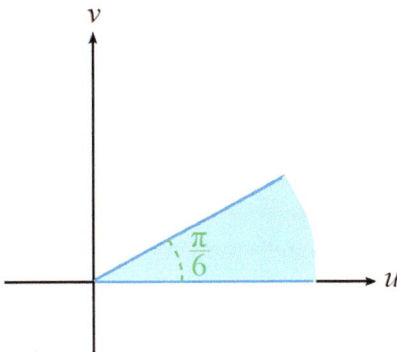

Fig. 4.3 The region in w-plane after the second complex transformation

Now, the transformed region needs to be transformed by $w_3 = (w_2)^6$ that will result in the region shown in Fig. 4.4.

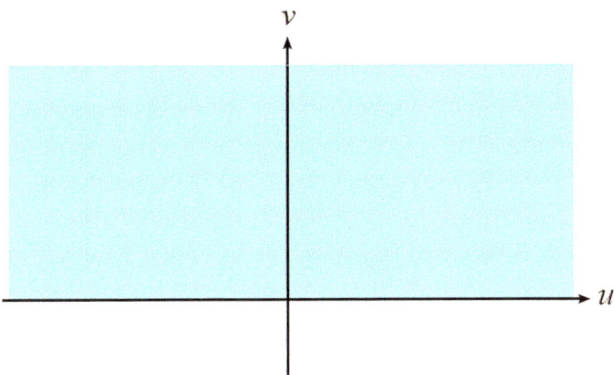

Fig. 4.4 The region in w-plane after the third complex transformation

Hence, the overall transformation is as follows:

$$w = (w_2)^6 = \left(e^{-i\frac{\pi}{6}} w_1\right)^6 = \left(e^{-i\frac{\pi}{6}}[z - (1+i)]\right)^6$$

$$\Rightarrow w = e^{-i\pi}[z - (1+i)]^6 = (\cos\pi - i\sin\pi)[z - (1+i)]^6$$

$$\Rightarrow w = -[z - (1+i)]^6$$

Choice (4) is the answer.

Notes

In this problem, the relations below have been used:

$$e^{i\theta} = \cos\theta + i\sin\theta$$

$$\cos = \pi - 1$$

$$\sin\pi = 0$$

4.2   Power Complex Transformation

4.3. Based on the information given in the problem, the transformation and the region in z-plane under transformation are as follows:

$$w = z^2 \tag{1}$$

$$\{x > 0, \quad y > 0, \quad xy > 1\} \tag{2}$$

As we know, $z = x + iy$ and $w = u + iv$. Therefore:

$$u + iv = (x + iy)^2 \tag{3}$$

$$\Rightarrow u + iv = x^2 - y^2 + i2xy \Rightarrow \begin{cases} u = x^2 - y^2 & (4) \\ v = 2xy & (5) \end{cases}$$

From (2), we can conclude that:

$$2xy > 2 \tag{6}$$

Solving (5) and (6):

$$v > 2$$

Choice (2) is the answer.

4.4. Based on the information given in the problem, the transformation and the region in z-plane under transformation are as follows:

$$w = z^2 \tag{1}$$

$$x^2 + y^2 = a^2 \tag{2}$$

As can be noticed from (2), the region under transformation is a circle with radius a and the center at the origin. Moreover, as we know, the equation of such a circle in polar coordinates is $|z| = a$. Also, the image of this circle under the transformation $w = z^n$ is a circle with the radius a^n and the center at the origin.

Therefore, the image of the circle (presented in (2)) under the transformation (presented in (1)) is a circle with the radius a^2 and the center at the origin. The equation of this circle in w-plane in Cartesian coordinates is as follows:

$$u^2 + v^2 = a^4$$

Choice (2) is the answer.

Notes

In this problem, the relation below has been used:

The equation of a circle with the radius R and the center at (x_0, y_0) is as follows:

$$(x - x_0)^2 + (y - y_0)^2 = R^2$$

4.5. Based on the information given in the problem, the transformation and the region in z-plane under transformation are as follows:

$$w = z^2 \tag{1}$$

$$Re(z) = 1 \Rightarrow x = 1 \tag{2}$$

As we know, $z = x + iy$ and $w = u + iv$. Therefore:

$$u + iv = (x + iy)^2 = x^2 - y^2 + i2xy \tag{3}$$

Solving (2) and (3):

$$u + iv = 1 - y^2 + i2y \Rightarrow \begin{cases} u = 1 - y^2 \\ v = 2y \end{cases}$$

Choice (4) is the answer.

4.6. Based on the information given in the problem, the transformation and the region in z-plane under transformation are as follows:

$$w = z^{\frac{1}{3}} \tag{1}$$

$$0 \le \theta \le \pi \tag{2}$$

As we know, the image of the region $\theta_1 \le \theta \le \theta_2$ under the transformation $w = z^n$ is the region $n\theta_1 \le \theta \le n\theta_2$.

Therefore, the image of the region (presented in (2)) under the transformation (presented in (1)) is $0 \le \theta \le \frac{\pi}{3}$ as shown in Fig. 4.5.

Choice (3) is the answer.

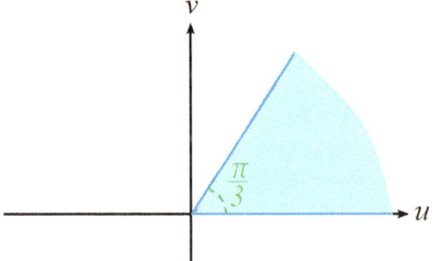

Fig. 4.5 The region in w-plane after complex transformation

4.7. Based on the information given in the problem, the region in z-plane under transformation (shown in Fig. 4.6) and the transformation are as follows:

$$r \leq 1, 0 \leq \theta \leq \frac{\pi}{2} \tag{1}$$

$$w = z^4 \tag{2}$$

As we know, the image of the region $\theta_1 \leq \theta \leq \theta_2$ under the transformation $w = z^n$ is the region $n\theta_1 \leq \theta \leq n\theta_2$. Therefore, the image of the first quadrant of the circle (presented in (1)) under the transformation (presented in (2)) is $r \leq 1$, $0 \leq \theta \leq 2\pi$ as shown in Fig. 4.7.

Choice (4) is the answer.

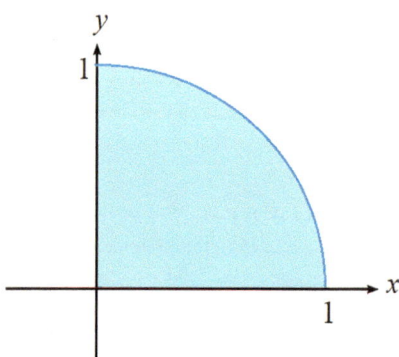

Fig. 4.6 The region in x-plane before the complex transformation

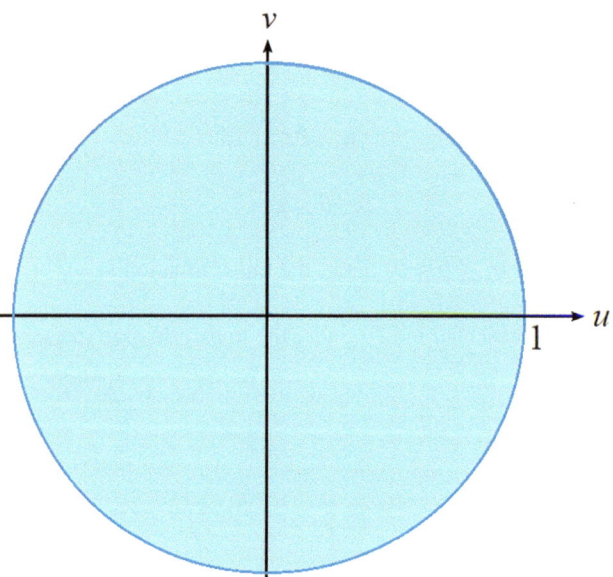

Fig. 4.7 The region in w-plane after the complex transformation

4.3 Reciprocal Complex Transformation

4.8. Based on the information given in the problem, the point (x, y) in z-plane needs to be transferred by the transformation below:

$$w = \frac{1}{z}$$

As we know, $z = x + iy$ and $w = u + iv$. Therefore:

$$u + iv = \frac{1}{x + iy}$$

$$\Rightarrow u + iv = \frac{1}{x + iy} \times \frac{x - iy}{x - iy} = \frac{x - iy}{x^2 + y^2}$$

$$\Rightarrow u + iv = \frac{x}{x^2 + y^2} - i\frac{y}{x^2 + y^2} \Rightarrow \begin{cases} u = \dfrac{x}{x^2 + y^2} \\ v = \dfrac{-y}{x^2 + y^2} \end{cases}$$

Choice (1) is the answer.

4.9. Based on the information given in the problem, the transformation and the region in z-plane under transformation are as follows:

$$w = \frac{1}{z} \Rightarrow z = \frac{1}{w} \tag{1}$$

$$x = 1 \tag{2}$$

As we know, $z = x + iy$ and $w = u + iv$. Therefore:

$$x + iy = \frac{1}{u + iv} \tag{3}$$

$$\Rightarrow x + iy = \frac{1}{u + iv} \times \frac{u - iv}{u - iv} = \frac{u - iv}{u^2 + v^2} \tag{4}$$

$$\Rightarrow x + iy = \frac{u}{u^2 + v^2} - i\frac{v}{u^2 + v^2} \Rightarrow \begin{cases} x = \dfrac{u}{u^2 + v^2} & (5) \\ y = \dfrac{-v}{u^2 + v^2} & (6) \end{cases}$$

Solving (2) and (5):

$$1 = \frac{u}{u^2 + v^2} \Rightarrow u^2 + v^2 = u \tag{7}$$

$$\Rightarrow u^2 + v^2 - u + \frac{1}{4} - \frac{1}{4} = 0 \Rightarrow \left(u - \frac{1}{2}\right)^2 + v^2 = \frac{1}{4} \tag{8}$$

The relation (8) shows the equation of a circle with the radius $\frac{1}{2}$ and the center at $\left(\frac{1}{2}, 0\right)$. Choice (1) is the answer.

Notes

In this problem, the relation below has been used:

The equation of a circle with the radius R and the center at (x_0, y_0) is as follows:

$$(x - x_0)^2 + (y - y_0)^2 = R^2$$

4.10. Based on the information given in the problem, the transformation and the region in z-plane under transformation are as follows:

$$w = \frac{i}{z} \Rightarrow z = \frac{i}{w} \tag{1}$$

$$|z - 1| = 1 \tag{2}$$

Considering $z = x + iy$ in (2):

$$|(x + iy) - 1| = 1 \Rightarrow |(x - 1) + iy| = 1 \tag{3}$$

$$\Rightarrow \sqrt{(x - 1)^2 + y^2} = 1 \Rightarrow (x - 1)^2 + y^2 = 1 \tag{4}$$

$$\Rightarrow x^2 - 2x + y^2 = 0 \tag{5}$$

Considering $z = x + iy$ and $w = u + iv$ in (1):

$$x + iy = \frac{i}{u + iv} \tag{6}$$

$$\Rightarrow x + iy = \frac{i}{u + iv} \times \frac{u - iv}{u - iv} = \frac{v + iu}{u^2 + v^2} \tag{7}$$

$$\Rightarrow x + iy = \frac{v}{u^2 + v^2} + i\frac{u}{u^2 + v^2} \Rightarrow \begin{cases} x = \dfrac{v}{u^2 + v^2} & (8) \\[2mm] y = \dfrac{u}{u^2 + v^2} & (9) \end{cases}$$

Solving (5), (8), and (9):

$$\left(\frac{v}{u^2 + v^2}\right)^2 - 2\left(\frac{v}{u^2 + v^2}\right) + \left(\frac{u}{u^2 + v^2}\right)^2 = 0 \tag{10}$$

$$\Rightarrow \frac{v^2 - 2v(u^2 + v^2) + u^2}{(u^2 + v^2)^2} = 0 \Rightarrow (u^2 + v^2)(1 - 2v) = 0$$

$$\Rightarrow v = \frac{1}{2}$$

Choice (2) is the answer.

Notes

In this problem, the relations below have been used:

$$|a + ib| = \sqrt{a^2 + b^2}$$

4.11. Based on the information given in the problem, the region in z-plane under transformation (shown in Fig. 4.8) and the transformation are as follows:

$$1 \leq y \leq 2 \tag{1}$$

$$w = \frac{1}{z} \tag{2}$$

As we know, $z = x + iy$ and $w = u + iv$. Therefore:

$$x + iy = \frac{1}{u + iv} \tag{3}$$

$$\Rightarrow x + iy = \frac{1}{u + iv} \times \frac{u - iv}{u - iv} = \frac{u - iv}{u^2 + v^2} \tag{4}$$

$$\Rightarrow x + iy = \frac{u}{u^2 + v^2} - i\frac{v}{u^2 + v^2} \Rightarrow \begin{cases} x = \dfrac{u}{u^2 + v^2} & (5) \\[3mm] y = \dfrac{-v}{u^2 + v^2} & (6) \end{cases}$$

Solving (1) and (6):

$$1 \leq \frac{-v}{u^2 + v^2} \leq 2 \Rightarrow \begin{cases} 1 \leq \dfrac{-v}{u^2 + v^2} & (7) \\[3mm] \dfrac{-v}{u^2 + v^2} \leq 2 & (8) \end{cases}$$

$$\Rightarrow \begin{cases} u^2 + v^2 + v \leq 0 \\[2mm] u^2 + v^2 + \dfrac{1}{2}v \geq 0 \end{cases} \Rightarrow \begin{cases} u^2 + \left(v + \dfrac{1}{2}\right)^2 \leq \dfrac{1}{4} & (9) \\[3mm] u^2 + \left(v + \dfrac{1}{4}\right)^2 \geq \dfrac{1}{16} & (10) \end{cases}$$

The relation (9) shows the inner surface area of a circle with the radius $\frac{1}{2}$ and the center at $\left(0, -\frac{1}{2}\right)$. Moreover, the relation (10) shows the outer surface area of a circle with the radius $\frac{1}{4}$ and the center at $\left(0, -\frac{1}{4}\right)$. The region in w-plane after the complex transformation is shown in Fig. 4.9. Choice (2) is the answer.

Notes

In this problem, the relations below have been used:

The equation of a circle with the radius R and the center at (x_0, y_0) is as follows:

$$(x - x_0)^2 + (y - y_0)^2 = R^2$$

The equation of the inner surface area of a circle with the radius R and the center at (x_0, y_0) is as follows:

$$(x - x_0)^2 + (y - y_0)^2 \leq R^2$$

The equation of the outer surface area of a circle with the radius R and the center at (x_0, y_0) is as follows:

$$(x - x_0)^2 + (y - y_0)^2 \geq R^2$$

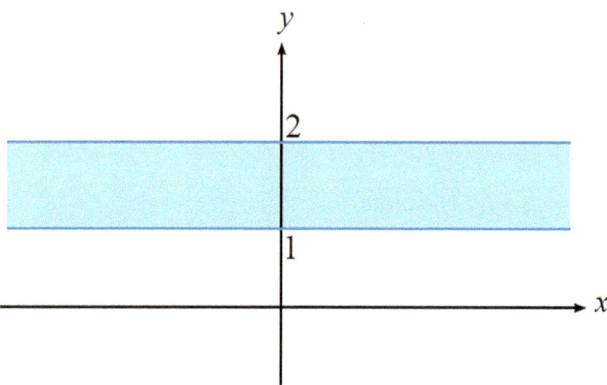

Fig. 4.8 The region in z-plane before the complex transformation

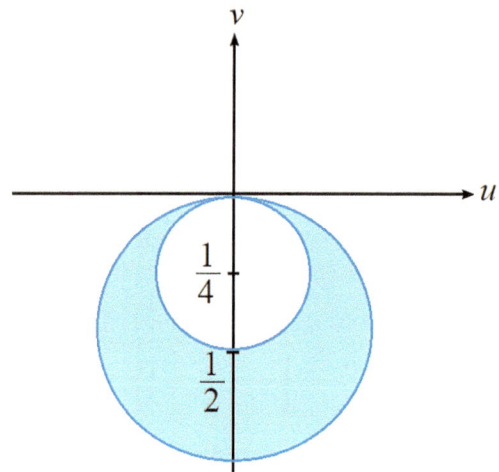

Fig. 4.9 The region in w-plane after the complex transformation

4.12. Based on the information given in the problem, the transformation and the region in z-plane under transformation are as follows:

$$w = u + iv = \frac{1}{z} \tag{1}$$

$$(x + 1)^2 + (y - 1)^2 = 2 \tag{2}$$

As we know, $z = x + iy$ and $w = u + iv$. Therefore:

$$x + iy = \frac{1}{u + iv} \tag{3}$$

$$\Rightarrow x + iy = \frac{1}{u + iv} \times \frac{u - iv}{u - iv} = \frac{u - iv}{u^2 + v^2} \tag{4}$$

$$\Rightarrow x + iy = \frac{u}{u^2 + v^2} - i\frac{v}{u^2 + v^2} \Rightarrow \begin{cases} x = \dfrac{u}{u^2 + v^2} \\[2mm] y = \dfrac{-v}{u^2 + v^2} \end{cases} \begin{array}{c} (5) \\[4mm] (6) \end{array}$$

From (2), we can write:

$$x^2 + 2x + 1 + y^2 - 2y + 1 = 2 \Rightarrow x^2 + 2x + y^2 - 2y = 0 \tag{7}$$

Solving (5, 6, and 7):

$$\left(\frac{u}{u^2 + v^2}\right)^2 + \frac{2u}{u^2 + v^2} + \left(\frac{-v}{u^2 + v^2}\right)^2 + \frac{2v}{u^2 + v^2} = 0 \tag{8}$$

$$\Rightarrow \frac{u^2 + 2u(u^2 + v^2) + v^2 + 2v(u^2 + v^2)}{(u^2 + v^2)^2} = 0 \Rightarrow (u^2 + v^2)(1 + 2u + 2v) = 0 \tag{9}$$

$$\Rightarrow 1 + 2u + 2v = 0 \Rightarrow u + v + \frac{1}{2} = 0$$

Choice (2) is the answer.

4.13. Based on the information given in the problem, the transformation and the region in z-plane under transformation are as follows:

$$w = \frac{1}{z} \tag{1}$$

$$2y + 3x - 1 = 0 \tag{2}$$

As we know, $z = x + iy$ and $w = u + iv$. Therefore:

$$x + iy = \frac{1}{u + iv} \tag{3}$$

$$\Rightarrow x + iy = \frac{1}{u + iv} \times \frac{u - iv}{u - iv} = \frac{u - iv}{u^2 + v^2} \tag{4}$$

$$\Rightarrow x + iy = \frac{u}{u^2 + v^2} - i\frac{v}{u^2 + v^2} \Rightarrow \begin{cases} x = \dfrac{u}{u^2 + v^2} \\[2mm] y = \dfrac{-v}{u^2 + v^2} \end{cases} \begin{array}{c} (5) \\[4mm] (6) \end{array}$$

Solving (2), (5), and (6):

$$\frac{-2v}{u^2 + v^2} + \frac{3u}{u^2 + v^2} - 1 = 0 \tag{7}$$

$$\Rightarrow u^2 + v^2 - 3u + 2v = 0 \tag{8}$$

$$\Rightarrow u^2 - 3u + \frac{9}{4} - \frac{9}{4} + v^2 + 2v + 1 - 1 = 0 \tag{9}$$

$$\Rightarrow \left(u - \frac{3}{2}\right)^2 + (v + 1)^2 = \frac{13}{4} \tag{10}$$

The relation (10) shows the equation of a circle with the radius $\frac{\sqrt{13}}{2}$ and the center at $\left(\frac{3}{2}, -1\right)$. Moreover, the point (0, 0) satisfies the equation of the circle as can be seen in the following:

$$\left(0 - \frac{3}{2}\right)^2 + (0 + 1)^2 = \frac{13}{4} \Rightarrow \frac{13}{4} = \frac{13}{4}$$

Thus, the circle passes from the origin. Choice (2) is the answer.

Notes

In this problem, the relation below has been used:

The equation of a circle with the radius R and the center at (x_0, y_0) is as follows:

$$(x - x_0)^2 + (y - y_0)^2 = R^2$$

4.14. Based on the information given in the problem, the transformation and the region in z-plane under transformation are as follows:

$$w = z + \frac{1}{z} \tag{1}$$

$$z = e^{i\theta}, \quad 0 \le \theta \le 2\pi \tag{2}$$

As we know, $w = u + iv$. Solving (1) and (2) results in the following relation:

$$u + iv = e^{i\theta} + \frac{1}{e^{i\theta}} \tag{3}$$

$$\Rightarrow u + iv = e^{i\theta} + e^{-i\theta} \tag{4}$$

$$\Rightarrow u + iv = 2\cos\theta \Rightarrow \begin{cases} u = 2\cos\theta \\ v = 0 \end{cases} \tag{5}$$

From (2), we can conclude that:

$$-1 \le \cos\theta \le 1 \Rightarrow -2 \le 2\cos\theta \le 2 \tag{6}$$

Solving (5) and (6):

$$-2 \leq u \leq 2$$

Choice (4) is the answer.

Notes

In this problem, the relations below have been used:

$$e^{i\theta} = \cos\theta + i\sin\theta$$

$$e^{-i\theta} = \cos\theta - i\sin\theta$$

4.4 Exponential Complex Transformation

4.15. Based on the information given in the problem, the transformation and the region in z-plane under transformation are as follows:

$$w = e^z \tag{1}$$

$$1 \leq x \leq 2, \quad \frac{\pi}{4} \leq y \leq \frac{\pi}{2} \tag{2}$$

From (1) and $z = x + iy$, it is concluded that:

$$w = e^{x+iy} = e^x e^{iy} \Rightarrow \begin{cases} |w| = e^x & (3) \\ \text{Arg}(w) = y & (4) \end{cases}$$

Solving (2) and (3):

$$e^1 \leq e^x \leq e^2 \Rightarrow e^1 \leq |w| \leq e^2 \tag{5}$$

Solving (2) and (4):

$$\frac{\pi}{4} \leq \text{Arg}(w) \leq \frac{\pi}{2} \tag{6}$$

As we know, $|w| = R$ is the equation of a circle with the radius R and the center at the origin. Thus, (5) is concerned with the region between two concentric circles with radii e^1 and e^2 and the center at the origin. Moreover, (6) is related to the region that their angle is between $\frac{\pi}{4}$ and $\frac{\pi}{2}$. The solution to the problem is graphically shown in Fig. 4.10. Choice (1) is the answer.

Notes

In this problem, the relations below have been used:

$$e^{a+b} = e^a e^b$$

$$\left|e^{x+iy}\right| = e^x$$

$$\mathrm{Arg}\left(e^{x+iy}\right) = y$$

The equation of a circle with radius R centered at z_0 is as follows:

$$|z - z_0| = R$$

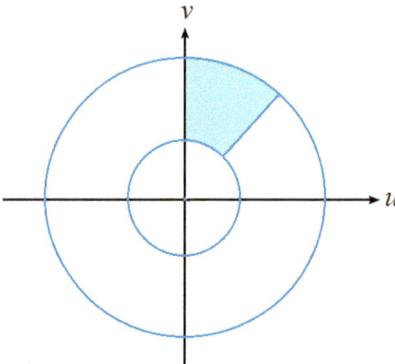

Fig. 4.10 The region in w-plane after the complex transformation

4.16. Based on the information given in the problem, the transformation and the region in z-plane under transformation are as follows:

$$w = e^z \qquad (1)$$

$$-\infty < x \leq 0, \quad 0 \leq y \leq \pi \qquad (2)$$

From (1) and $z = x + iy$, it is concluded that:

$$w = e^{x+iy} = e^x e^{iy} \Rightarrow \begin{cases} |w| = e^x & (3) \\ \mathrm{Arg}(w) = y & (4) \end{cases}$$

Solving (2) and (3):

$$e^{-\infty} < e^x \leq e^0 \Rightarrow 0 < |w| \leq 1 \qquad (5)$$

Solving (2) and (4):

$$0 \leq \mathrm{Arg}(w) \leq \pi \qquad (6)$$

As we know, $|w| = R$ is the equation of a circle with the radius R and the center at the origin. Thus, (5) is concerned with the inner region of a circle with the radius 1 and the center at the origin. Moreover, (6) is related to the region that their angle is between 0 and π. Hence, the solution is the upper half of a circle with the radius 1, which is graphically shown in Fig. 4.11. Choice (1) is the answer.

Notes

In this problem, the relations below have been used:

$$e^{a+b} = e^a e^b$$

$$\left| e^{x+iy} \right| = e^x$$

$$\text{Arg}\left(e^{x+iy} \right) = y$$

The equation of a circle with the radius R centered at z_0 is as follows:

$$|z - z_0| = R$$

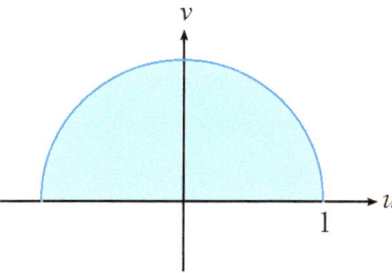

Fig. 4.11 The region in w-plane after the complex transformation

4.17. Based on the information given in the problem, the transformation and the region in z-plane under transformation are as follows:

$$w = e^z \tag{1}$$

$$Re\{z\} \leq 0, \quad 0 \leq Im\{z\} \leq \frac{\pi}{2} \implies -\infty < x \leq 0, \quad 0 \leq y \leq \frac{\pi}{2} \tag{2}$$

From (1) and $z = x + iy$, it is concluded that:

$$w = e^{x+iy} = e^x e^{iy} \implies \begin{cases} |w| = e^x & (3) \\ \text{Arg}(w) = y & (4) \end{cases}$$

Solving (2) and (3):

$$e^{-\infty} < e^x \leq e^0 \implies 0 < |w| \leq 1 \tag{5}$$

Solving (2) and (4):

$$0 \le \mathrm{Arg}(w) \le \frac{\pi}{2} \tag{6}$$

As we know, $|w| = R$ is the equation of a circle with the radius R and the center at the origin. Thus, (5) is concerned with the inner region of a circle with the radius 1 and the center at the origin. Moreover, (6) is related to the region that their angle is between 0 and $\frac{\pi}{2}$. Hence, the solution is the first quadrant of the unit circle, which is graphically shown in Fig. 4.12. Choice (2) is the answer.

Notes

In this problem, the relations below have been used:

$$e^{a+b} = e^a e^b$$

$$\left| e^{x+iy} \right| = e^x$$

$$\mathrm{Arg}\left(e^{x+iy} \right) = y$$

The equation of a circle with the radius R centered at z_0 is as follows:

$$|z - z_0| = R$$

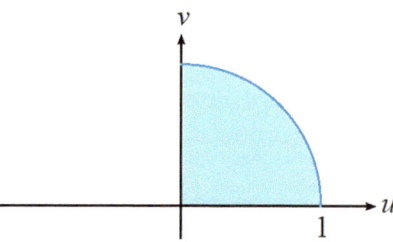

Fig. 4.12   The region in w-plane after the complex transformation

4.5   Natural Logarithm Complex Transformation

4.18. Based on the information given in the problem, the transformation and the region in z-plane under transformation are as follows:

$$w = \ln(z - 1) \tag{1}$$

$$1 < x < \infty, \quad 0 \le y < \infty \tag{2}$$

From (1) and $z = x + iy$ and $w = u + iv$, it is concluded that:

$$u + iv = \ln[(x - 1) + iy] \tag{3}$$

$$\Rightarrow u + iv = \ln\left[\sqrt{(x-1)^2 + y^2}\, e^{i\tan^{-1}\frac{y}{x-1}}\right] \tag{4}$$

$$\Rightarrow u + iv = \ln\left[\sqrt{(x-1)^2 + y^2}\right] + i\tan^{-1}\frac{y}{x-1} \Rightarrow \begin{cases} u = \ln\left[\sqrt{(x-1)^2 + y^2}\right] & \text{(5)} \\ \\ v = \tan^{-1}\frac{y}{x-1} & \text{(6)} \end{cases}$$

Solving (2) and (5):

$$-\infty < \ln\left[\sqrt{(x-1)^2 + y^2}\right] < \infty \Rightarrow -\infty < u < \infty \tag{7}$$

Moreover, the value of $\frac{y}{x-1}$ for the range presented in (2) is as follows:

$$0 < \frac{y}{x-1} < \infty \tag{8}$$

Solving (6) and (8):

$$0 < \tan^{-1}\frac{y}{x-1} < \frac{\pi}{2} \Rightarrow 0 < v < \frac{\pi}{2} \tag{9}$$

The solution to the problem is graphically shown in Fig. 4.13. Choice (2) is the answer.

Notes

In this problem, the relations below have been used:

$$\begin{cases} a + ib = \sqrt{a^2 + b^2}\, e^{i\tan^{-1}\left|\frac{b}{a}\right|} & \text{if } a > 0, b > 0 \\ a + ib = \sqrt{a^2 + b^2}\, e^{i\left(\pi - \tan^{-1}\left|\frac{b}{a}\right|\right)} & \text{if } a < 0, b > 0 \\ a + ib = \sqrt{a^2 + b^2}\, e^{i\left(\pi + \tan^{-1}\left|\frac{b}{a}\right|\right)} & \text{if } a < 0, b < 0 \\ a + ib = \sqrt{a^2 + b^2}\, e^{-i\tan^{-1}\left|\frac{b}{a}\right|} & \text{if } a > 0, b < 0 \end{cases}$$

$$\ln|z|e^{i\theta_z} = \ln|z| + i\theta_z$$

$$\ln 0^+ = -\infty$$

$$\ln\infty = \infty$$

$$\tan^{-1}0^+ = 0^+$$

$$\tan^{-1}\infty = \frac{\pi}{2}$$

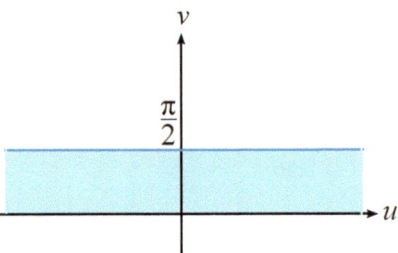

Fig. 4.13 The region in w-plane after the complex transformation

4.6   Hyperbolic Sine and Cosine Complex Transformation

4.19. Based on the information given in the problem, the transformation and the region in z-plane under transformation are as follows:

$$w = \cosh z \tag{1}$$

$$y = \frac{\pi}{4} \tag{2}$$

As we know, $z = x + iy$ and $w = u + iv$. Therefore:

$$u + iv = \cosh(x + iy) \tag{3}$$

$$\Rightarrow u + iv = \cosh x \cosh iy + \sinh x \sinh iy \tag{4}$$

$$\Rightarrow u + iv = \cosh x \cos y + i \sinh x \sin y \Rightarrow \begin{cases} u = \cosh x \cos y \\ v = \sinh x \sin y \end{cases} \tag{5}$$

Solving (2) and (5):

$$\begin{cases} u = \dfrac{\sqrt{2}}{2} \cosh x \\ v = \dfrac{\sqrt{2}}{2} \sinh x \end{cases} \Rightarrow \begin{cases} \cosh x = \dfrac{2u}{\sqrt{2}} \\ \sinh x = \dfrac{2v}{\sqrt{2}} \end{cases} \tag{6}$$

As we know:

$$\cosh^2 x - \sinh^2 x = 1 \tag{7}$$

Solving (6) and (7):

$$\frac{4u^2}{2} - \frac{4v^2}{2} = 1 \Rightarrow u^2 - v^2 = \frac{1}{2}$$

Choice (3) is the answer.

In this problem, the relations below have been used:

$$\cosh(a + b) = \cosh a \cosh b + \sinh a \sinh b$$

$$\cosh iy = \cos y$$

$$\sinh iy = i \sin y$$

$$\cos \frac{\pi}{4} = \frac{\sqrt{2}}{2}$$

$$\sin \frac{\pi}{4} = \frac{\sqrt{2}}{2}$$

4.7   Sine and Cosine Complex Transformation

4.20. Based on the information given in the problem, the transformation and the region in z-plane under transformation are as follows:

$$w = \sin z \tag{1}$$

$$x = \frac{\pi}{6} \tag{2}$$

As we know, $z = x + iy$ and $w = u + iv$. Therefore:

$$u + iv = \sin(x + iy) \tag{3}$$

$$\Rightarrow u + iv = \sin x \cos iy + \cos x \sin iy \tag{4}$$

$$\Rightarrow u + iv = \sin x \cosh y + i \cos x \sinh y \Rightarrow \begin{cases} u = \sin x \cosh y \\ v = \cos x \sinh y \end{cases} \tag{5}$$

Solving (2) and (5):

$$\begin{cases} u = \frac{1}{2} \cosh y \\ v = \frac{\sqrt{3}}{2} \sinh y \end{cases} \Rightarrow \begin{cases} \cosh y = 2u \\ \sinh y = \frac{2v}{\sqrt{3}} \end{cases} \tag{6}$$

As we know:

$$\cosh^2 y - \sin^2 y = 1 \tag{7}$$

Solving (6) and (7):

$$4u^2 - \frac{4v^2}{3} = 1 \Rightarrow \frac{u^2}{\frac{1}{4}} - \frac{v^2}{\frac{3}{4}} = 1 \tag{8}$$

The relation (8) is the equation of a hyperbola. Choice (4) is the answer.

In this problem, the relations below have been used:

$$\sin(a+b) = \sin a \cos b + \cos a \sin b$$

$$\cosh iy = \cos y$$

$$\sinh iy = i \sin y$$

$$\cos \frac{\pi}{6} = \frac{\sqrt{3}}{2}$$

$$\sin \frac{\pi}{6} = \frac{1}{2}$$

The general equation of hyperbola centered at $(-x_0, -y_0)$ is as follows:

$$\frac{(x-x_0)^2}{a^2} - \frac{(y-y_0)^2}{b^2} = 1$$

4.8   Linear Fractional Complex Transformation

4.21. Based on the information given in the problem, we need to find a linear fractional transformation that maps the three points $z_1 = 0$, $z_2 = -i$, and $z_3 = -1$ in z-plane to $w_1 = i$, $w_2 = 1$, and $w_3 = 0$ in w-plane. In other words:

$$\begin{cases} z_1 = 0 \rightarrow w_1 = i \\ z_1 = -i \rightarrow w_1 = 1 \\ z_1 = -1 \rightarrow w_1 = 0 \end{cases}$$

The problem can be solved as follows:

$$\frac{w - w_1}{w - w_3} \frac{w_2 - w_3}{w_2 - w_1} = \frac{z - z_1}{z - z_3} \frac{z_2 - z_3}{z_2 - z_1}$$

$$\Rightarrow \frac{w-i}{w-0} \frac{1-0}{1-i} = \frac{z-0}{z+1} \frac{-i+1}{-i-0} \Rightarrow \frac{w-i}{w} \frac{1}{1-i} = \frac{z}{z+1} \frac{1-i}{-i}$$

$$\Rightarrow \frac{w-i}{w} = \frac{z}{z+1}(1-i)^2 \frac{1}{-i} \Rightarrow \frac{w-i}{w} = \frac{z}{z+1}(1-2i-1)\frac{1}{-i}$$

$$\Rightarrow \frac{w-i}{w} = \frac{2z}{z+1} \Rightarrow 1 - \frac{i}{w} = \frac{2z}{z+1}$$

$$\Rightarrow \frac{i}{w} = 1 - \frac{2z}{z+1} \Rightarrow \frac{i}{w} = \frac{1-z}{z+1}$$

$$\Rightarrow w = \frac{z+1}{1-z}i \Rightarrow w = -i\left(\frac{z+1}{z-1}\right)$$

Choice (2) is the answer.

Notes

In this problem, the relations below have been used:

$$(a + b)^2 = a^2 + 2ab + b^2$$

$$i^2 = -1$$

References

1. Rahmani-Andebili, M. (2024). Precalculus (2nd Ed.) – Practice Problems, Methods, and Solutions, Springer Nature.
2. Rahmani-Andebili, M. (2023). Calculus III – Practice Problems, Methods, and Solutions, Springer Nature.
3. Rahmani-Andebili, M. (2023). Calculus II – Practice Problems, Methods, and Solutions, Springer Nature.
4. Rahmani-Andebili, M. (2023). Calculus I (2nd Ed.) – Practice Problems, Methods, and Solutions, Springer Nature.
5. Rahmani-Andebili, M. (2022). Differential Equations – Practice Problems, Methods, and Solutions, Springer Nature.
6. Rahmani-Andebili, M. (2021). Calculus – Practice Problems, Methods, and Solutions, Springer Nature.
7. Rahmani-Andebili, M. (2021). Precalculus – Practice Problems, Methods, and Solutions, Springer Nature.

Singularities of Complex Functions, Complex series, Taylor and Laurent Series Expansions of Complex Functions, and Residue of Complex Functions: Problems

5

Abstract

In this chapter, the basic and advanced problems concerned with the singularities of complex functions including poles, removable singularity, and essential singularity; complex series; Taylor and Laurent series expansions of complex functions; and residue of complex functions are presented and studied. Herein, different types of problems and exercises are presented that are categorized as follows:

○ *Problems with detailed solution*: They have been designed to teach students the subjects in detail. Moreover, they have been categorized into different levels based on their difficulty levels (easy, normal, and hard) and calculation amounts (small, normal, and large).

○ *Partially solved exercises*: They have been designed to encourage students to practice more problems while guiding them through the problem-solving procedure and hinting the required formulas.

○ *Exercises with final answer*: They have been designed to encourage students to practice by themselves while hinting them by the final answer as well as to help instructors to give tests or quizzes.

5.1 Poles, Removable Singularity, and Essential Singularity of a Complex Function

5.1. Determine the poles of the complex function below [1–7]:

$$f(z) = \frac{z}{\sinh z \cosh z}$$

Difficulty level ○ Easy ○ Normal ● Hard
Calculation amount ● Small ○ Normal ○ Large

1) $i\frac{k\pi}{2}, k\epsilon\mathbb{Z} - \{0\}$
2) $i\frac{k\pi}{2}, k\epsilon\mathbb{Z}$
3) $ik\pi, k\epsilon\mathbb{Z} - \{0\}$
4) $ik\pi, k\epsilon\mathbb{Z}$

Exercise

Determine the number of poles of the complex function below in the region $|z| < 2$:

$$f(z) = \frac{z}{\sin z - 1}$$

Final Answer
One

5.2. What is the type of the point $z = 0$ for the complex function below:

$$f(z) = \frac{1 - \cos z}{z^3}$$

Difficulty level   ○ Easy   ● Normal   ○ Hard
Calculation amount   ● Small   ○ Normal   ○ Large
1) An essential singular point
2) A first-order pole
3) A removable singular point
4) A second-order pole

5.3. Determine the order of the pole $z = 0$ of the complex function below:

$$f(z) = \frac{\cos z - 1}{z^4}$$

Difficulty level   ○ Easy   ○ Normal   ● Hard
Calculation amount   ● Small   ○ Normal   ○ Large
1) 1
2) 2
3) 3
4) The function does not have any pole.

Exercise

Determine the order of the pole $z = 0$ for the complex function below:

$$f(z) = \frac{\sin z}{z^4}$$

Final Answer
Three

5.4. Determine the number of singular points of the complex function below in the region $|z| < 2$:

$$f(z) = \frac{\ln(z+3)}{(z^2+2)\sin z}$$

Difficulty level   ○ Easy   ● Normal   ○ Hard
Calculation amount   ● Small   ○ Normal   ○ Large
1) 1
2) 2
3) 3
4) ∞

Exercise

Determine the number of singular points of the complex function below in the region $|z| < 5$:

$$f(z) = \frac{1}{(z^2+24)\sin z}$$

Final Answer

Five

5.5. What is the type of the point $z = 0$ for the complex function below:

$$f(z) = z^2 e^{\frac{1}{z}}$$

Difficulty level   ○ Easy   ● Normal   ○ Hard
Calculation amount   ● Small   ○ Normal   ○ Large
1) An essential singular point
2) A first-order pole
3) A removable singular point
4) A second-order pole

Partially Solved Exercise

What is the type of the point $z = 0$ for the complex function below:

$$f(z) = z^{10} e^{-\frac{1}{z^2}}$$

Solution

The Laurent series expansion of the exponential term can be used in the main function as follows:

$$f(z) = z^{10}(\qquad\qquad\qquad\qquad\qquad)$$

$$\Rightarrow f(z) =$$

Since the Laurent series expansion of the complex function includes an infinite number of terms with a negative exponent of z, the point $z = 0$ for the function is an essential singular point.

Notes

In this problem, the relations below have been used:

$$e^z = \sum_{n=0}^{\infty} \frac{z^n}{n!} = 1 + z + \frac{z^2}{2!} + \cdots$$

If the Laurent series expansion of a complex function includes an infinite number of terms with a negative exponent of $z - z_0$, the point $z = z_0$ for the function is an essential singular point. Moreover, if the Laurent series expansion includes a finite number of terms with a negative exponent of $z - z_0$, the point $z = z_0$ is called a pole. However, if the point $z = z_0$ is the root of both numerator and denominator, and the limit of the function at this point is finite, the point $z = z_0$ is called a removable singular point.

5.6. What is the type of the point $z = 0$ for the complex function below?

$$f(z) = \frac{e^z - 1}{z^2}$$

Difficulty level   ○ Easy   ○ Normal   ● Hard
Calculation amount   ○ Small   ● Normal   ○ Large
1) An essential singular point
2) A first-order pole
3) A removable singular point
4) A second-order pole

5.7. Which one of the following choices is correct about the complex function below at $z = 1$?

$$f(z) = e^{-\frac{1}{(z-1)^2}}$$

Difficulty level   ○ Easy   ● Normal   ○ Hard
Calculation amount   ● Small   ○ Normal   ○ Large
1) $z = 1$ is a second-order pole.
2) $z = 1$ is an essential singular point.
3) $z = 1$ is a removable singular point.
4) $z = 1$ is a first-order pole.

5.8. Determine the singular points of the following complex function:

$$f(z) = \cot(\pi z)$$

Difficulty level   ● Easy   ○ Normal   ○ Hard
Calculation amount   ● Small   ○ Normal   ○ Large

1) 0

2) $0, \pm \dfrac{1}{\pi}, \pm \dfrac{2}{\pi},$

3) $0, \pm 1, \pm 2, \cdots$

4) $0, \pm 1, \pm \dfrac{1}{2},$

Exercise

Determine the singular points of the following complex function:

$$f(z) = \tan(z)$$

Final Answer

$0, \pm \frac{\pi}{2}, \pm \frac{3\pi}{2}, \cdots$

5.2 Complex Series

5.9. Calculate the convergence radius of the complex power series below:

$$\sum_{n=1}^{\infty} \left(1 - \frac{1}{n}\right)^{n^2} z^n$$

Difficulty level   ○ Easy   ○ Normal   ● Hard

Calculation amount   ● Small   ○ Normal   ○ Large

1) e^{-1}

2) e

3) 0

4) 1

Exercise

Determine the convergence radius of the complex power series below:

$$\sum_{n=1}^{\infty} \left(1 - \frac{2}{n}\right)^{3n^2} z^n$$

Final Answer

e^6

5.10. Calculate the convergence radius of the following complex power series:

$$\sum_{n=0}^{\infty} \frac{(2n)!}{(n!)^2} z^n$$

1) $\dfrac{1}{2}$

2) $\dfrac{1}{4}$

3) 4

4) 2

5.11. Calculate the convergence radius of the following complex power series:

$$\sum_{n=0}^{\infty} (\cos in) z^n$$

1) e^{-1}

2) e^{-2}

3) $2e^{-1}$

4) e

5.12. Calculate the convergence region of the complex series below:

$$f(z) = \sum_{n=1}^{\infty} e^{\frac{i}{z+1}n}$$

1) $x > -1$

2) $y < 0$

3) $y > 0$

4) $x + 1 > y$

Partially Solved Exercise

Calculate the convergence region of the complex function below:

$$f(z) = \sum_{n=1}^{\infty} \left(\frac{2z}{2z+1} \right)^n$$

Solution

The criterion for the convergence of the series is as follows:

$$\left| \frac{2z}{2z+1} \right| < 1$$

$$\Rightarrow |2z| < |2z + 1|$$

$$\Rightarrow |2(x + iy)| < |2(x + iy) + 1|$$

$$\Rightarrow |(\quad) + i(\quad)| < |(\quad) + i(\quad)|$$

$$\Rightarrow (\quad) < (\quad)$$

$$\Rightarrow (\quad) < (\quad)$$

$$\Rightarrow x > -\frac{1}{4}$$

Notes

In this problem, the relations below have been used:

$$z = x + iy$$

$$|a + ib| = \sqrt{a^2 + b^2}$$

5.3 Taylor and Laurent Series Expansions of Complex Functions

5.13. Calculate the Laurent series expansion of the complex function below:

$$f(z) = \ln\left(\frac{1 + z}{1 - z}\right)$$

Difficulty level   ○ Easy   ○ Normal   ● Hard
Calculation amount   ○ Small   ○ Normal   ● Large

1) $\sum\limits_{n=0}^{\infty} \frac{3z^{2n+1}}{2n+1}$

2) $\sum\limits_{n=0}^{\infty} \frac{4z^{2n+1}}{2n+1}$

3) $\sum\limits_{n=0}^{\infty} \frac{z^{2n+1}}{2n+1}$

4) $\sum\limits_{n=0}^{\infty} \frac{2z^{2n+1}}{2n+1}$

5.14. Determine the Laurent series expansion of the complex function $f(z) = \tan^{-1} z$ if $|z| < 1$.
Difficulty level   ○ Easy   ○ Normal   ● Hard
Calculation amount   ○ Small   ○ Normal   ● Large

1) $z + \frac{z^3}{3!} - \frac{z^5}{5!} +$

2) $z + \frac{z^3}{3} - \frac{z^5}{5} +$

3) $z - \frac{z^3}{3!} + \frac{z^5}{5!} -$

4) $z - \frac{z^3}{3} + \frac{z^5}{5} -$

5.15. Calculate the Laurent series expansion of the complex function below for the region described by $1 < |z - 1| < 2$:

$$f(z) = \frac{z}{(z-1)(z-3)}$$

Difficulty level   ○ Easy   ○ Normal   ● Hard
Calculation amount   ○ Small   ○ Normal   ● Large

1) $-\frac{1}{2(z-1)} - \frac{3}{4} \sum\limits_{n=0}^{\infty} \left(\frac{z-1}{2}\right)^n$

2) $\frac{1}{2(z-1)} - \frac{3}{4} \sum\limits_{n=0}^{\infty} \left(\frac{z-1}{2}\right)^n$

3) $-\frac{1}{2(z-1)} + \frac{3}{4} \sum\limits_{n=0}^{\infty} \left(\frac{z-1}{2}\right)^n$

4) $\frac{1}{2(z-1)} + \frac{3}{4} \sum\limits_{n=0}^{\infty} \left(\frac{z-1}{2}\right)^n$

5.16. Calculate the Laurent series expansion of the complex function below in the region $|z| > 4$:

$$f(z) = \frac{1}{z-4}$$

Difficulty level   ○ Easy   ○ Normal   ● Hard
Calculation amount   ○ Small   ● Normal   ○ Large

1) $\sum\limits_{n=1}^{\infty} \frac{4^n}{z^{n-1}}$

2) $\sum\limits_{n=1}^{\infty} \left(\frac{4}{z}\right)^{n+1}$

3) $\sum\limits_{n=0}^{\infty} \frac{4^{n+1}}{z^n}$

4) $\sum\limits_{n=1}^{\infty} \frac{4^{n-1}}{z^n}$

5.4   Residue of Complex Functions

5.17. Determine the residue of the following complex function at point $z = 1$:

$$f(z) = \frac{z}{(z-1)(z+1)^2}$$

Difficulty level   ● Easy   ○ Normal   ○ Hard
Calculation amount   ● Small   ○ Normal   ○ Large

1) $-\frac{1}{4}$

2) $\frac{1}{4}$

3) $\frac{1}{2}$

4) $-\frac{1}{2}$

Partially Solved Exercise

Determine the residue of the following complex function at point $z = -1$:

$$f(z) = \frac{z}{(z-1)(z+1)^2}$$

Solution

The residue of a function $f(z)$ at $z = z_0$ can be calculated by using two methods as follows:

Method 1: The coefficient of $\frac{1}{z - z_0}$ in the Laurent series expansion of a function $f(z)$ at $z = z_0$ is the residue.

Method 2: If z_0 is the mth-order pole of a complex function $f(z)$, then the residue of the function at $z = z_0$ can be calculated as follows:

$$\underset{z = z_0}{\mathrm{Res}} f(z) = \frac{1}{(m-1)!} \lim_{z \to z_0} \frac{d^{m-1}}{dz^{m-1}} [(z - z_0)^m f(z)]$$

If z_0 is the first-order pole of a complex function $f(z)$, then the residue of the function at $z = z_0$ can be calculated as follows:

$$\underset{z = z_0}{\mathrm{Res}} f(z) = \lim_{z \to z_0} (z - z_0) f(z)$$

Therefore, for this problem, we can write:

$$\underset{z = -1}{\mathrm{Res}} f(z) = \frac{1}{(2-1)!} \lim_{z \to -1} \left[\frac{d}{dz} (\qquad) \right] = \lim_{z \to -1} (\qquad)$$

$$\Rightarrow \underset{z = -1}{\mathrm{Res}} f(z) = -\frac{1}{4}$$

5.18. Calculate the residue of the complex function below at point $z = -2$:

$$f(z) = \frac{1}{z(z+2)^3}$$

Difficulty level   ○ Easy   ● Normal   ○ Hard
Calculation amount   ○ Small   ● Normal   ○ Large

1) $\frac{1}{4}$

2) $\frac{1}{8}$

3) $-\frac{1}{4}$

4) $-\frac{1}{8}$

Calculate the residue of the complex function below at point $z = 0$:

$$f(z) = \frac{1}{z(z+2)^3}$$

Final Answer

$\frac{1}{8}$

5.19. Calculate the residue of the complex function below at point $z = -1$:

$$f(z) = \frac{e^z}{(z+1)^3(z-2)}$$

Difficulty level   ○ Easy   ● Normal   ○ Hard
Calculation amount   ○ Small   ○ Normal   ● Large

1) $-\dfrac{1}{54}$

2) $-\dfrac{1}{54e}$

3) $\dfrac{1}{54}$

4) $\dfrac{1}{54e}$

Calculate the residue of the complex function below at point $z = 2$:

$$f(z) = \frac{e^z}{(z+1)^3(z-2)}$$

Final Answer

$\frac{e^2}{27}$

5.20. Calculate the residue of the complex function below at point $z = -2i$:

$$f(z) = \frac{z^2 - 2z}{(z+1)^2(z^2+4)}$$

Difficulty level   ○ Easy   ○ Normal   ● Hard
Calculation amount   ○ Small   ● Normal   ○ Large

1) $\dfrac{1+i}{3+4i}$

2) $\dfrac{1-i}{3+4i}$

3) $\dfrac{1-i}{3-4i}$

4) $\dfrac{1+i}{3-4i}$

Partially Solved Exercise

Calculate the residue of the complex function below at point $z = 1$:

$$f(z) = \frac{1}{1-z^4}$$

Solution

The residue of a function $f(z)$ at $z = z_0$ can be calculated by using two methods as follows:

Method 1: The coefficient of $\frac{1}{z-z_0}$ in the Laurent series expansion of a function $f(z)$ at $z = z_0$ is the residue.

Method 2: If z_0 is the mth-order pole of a complex function $f(z)$, then the residue of the function at $z = z_0$ can be calculated as follows:

$$\operatorname*{Res}_{z=z_0} f(z) = \frac{1}{(m-1)!} \lim_{z \to z_0} \frac{d^{m-1}}{dz^{m-1}} [(z-z_0)^m f(z)]$$

If z_0 is the first-order pole of a complex function $f(z)$, then the residue of the function at $z = z_0$ can be calculated as follows:

$$\operatorname*{Res}_{z=z_0} f(z) = \lim_{z \to z_0} (z-z_0) f(z)$$

Therefore:

$$\operatorname*{Res}_{z=1} f(z) = \lim_{z \to 1} \frac{(\quad\quad)}{(\quad\quad)}$$

$$\Rightarrow \operatorname*{Res}_{z=1} f(z) = \frac{0}{0}$$

$$\overset{H}{\Rightarrow} \operatorname*{Res}_{z=1} f(z) = \lim_{z \to 1} \frac{(\quad\quad)}{(\quad\quad)}$$

$$\Rightarrow \operatorname*{Res}_{z=1} f(z) = -\frac{1}{4}$$

Notes

In this problem, the relations below have been used:

$$h(z) = \lim_{z \to z_0} \frac{f(z)}{g(z)} = \frac{0}{0} \overset{H}{\Rightarrow} h(z) = \lim_{z \to z_0} \frac{f(z)}{g(z)}$$

5.21. Calculate the residue of the complex function below at point $z = 0$:

$$f(z) = z^2 e^{\frac{1}{z}}$$

Difficulty level   ○ Easy   ● Normal   ○ Hard
Calculation amount   ● Small   ○ Normal   ○ Large

1) 0

2) $\dfrac{1}{6}$

3) $-\dfrac{1}{6}$

4) $\dfrac{1}{3}$

5.22. Calculate the residue of the complex function below at point $z = 0$:

$$f(z) = \frac{1 - \cos z}{z^3}, \quad z \neq 0$$

Difficulty level   ○ Easy   ● Normal   ○ Hard
Calculation amount   ● Small   ○ Normal   ○ Large

1) $\dfrac{1}{2}$

2) 1

3) $\dfrac{1}{6}$

4) $\dfrac{1}{24}$

Partially Solved Exercise

Calculate the residue of the complex function below at point $z = 1$:

$$f(z) = (z - 1)^5 \cos \frac{1}{z - 1}$$

Solution

The residue of a function $f(z)$ at $z = z_0$ can be calculated by using two methods as follows:

Method 1: The coefficient of $\frac{1}{z - z_0}$ in the Laurent series expansion of a function $f(z)$ at $z = z_0$ is the residue.

Method 2: If z_0 is the mth-order pole of a complex function $f(z)$, then the residue of the function at $z = z_0$ can be calculated as follows:

$$\underset{z=z_0}{\mathrm{Res}} f(z) = \frac{1}{(m-1)!} \lim_{z \to z_0} \frac{d^{m-1}}{dz^{m-1}} [(z-z_0)^m f(z)]$$

If z_0 is the first-order pole of a complex function $f(z)$, then the residue of the function at $z = z_0$ can be calculated as follows:

$$\underset{z=z_0}{\mathrm{Res}} f(z) = \lim_{z \to z_0} (z-z_0) f(z)$$

The Laurent series expansion of the cosine term can be used in the main function as follows:

$$f(z) = (z-1)^5 (\qquad\qquad\qquad\qquad\qquad)$$

$$\Rightarrow f(z) =$$

Therefore, the residue of the function at point $z = 0$ is $-\frac{1}{6!}$.

Notes

In this problem, the relation below has been used:

$$\cos z = \sum_{n=0}^{\infty} (-1)^n \frac{z^{2n}}{(2n)!} = 1 - \frac{z^2}{2!} + \frac{z^4}{4!} - \cdots, \quad |z| < \infty$$

5.23. Determine the residue of the following complex function at point $z = 0$:

$$f(z) = \frac{\sinh z}{z^4}$$

Difficulty level   ○ Easy   ● Normal   ○ Hard
Calculation amount   ● Small   ○ Normal   ○ Large
1) 1
2) $\frac{1}{6}$
3) $\frac{1}{5!}$
4) $\frac{1}{24}$

5.24. Calculate the residue of the complex function below at point $z = 0$:

$$f(z) = z^7 \cos \frac{1}{z^2}$$

Difficulty level   ○ Easy   ● Normal   ○ Hard
Calculation amount   ● Small   ○ Normal   ○ Large
1) 0
2) $\frac{i}{12}$

3) $\dfrac{1}{24}$

4) $\dfrac{1}{8}$

5.25. Calculate the residue of the complex function below at point $z = \frac{3\pi}{2}$:

$$f(z) = e^z \tan z$$

Difficulty level   ○ Easy   ○ Normal   ● Hard
Calculation amount   ○ Small   ● Normal   ○ Large

1) $e^{\frac{3}{2}\pi}$

2) $e^{-\frac{3}{2}\pi}$

3) 3) $-e^{\frac{3}{2}\pi}$

4) 4) $-e^{-\frac{3}{2}\pi}$

References

1. Rahmani-Andebili, M. (2024). Precalculus (2nd Ed.) – Practice Problems, Methods, and Solutions, Springer Nature.
2. Rahmani-Andebili, M. (2023). Calculus III – Practice Problems, Methods, and Solutions, Springer Nature.
3. Rahmani-Andebili, M. (2023). Calculus II – Practice Problems, Methods, and Solutions, Springer Nature.
4. Rahmani-Andebili, M. (2023). Calculus I (2nd Ed.) – Practice Problems, Methods, and Solutions, Springer Nature.
5. Rahmani-Andebili, M. (2022). Differential Equations – Practice Problems, Methods, and Solutions, Springer Nature.
6. Rahmani-Andebili, M. (2021). Calculus – Practice Problems, Methods, and Solutions, Springer Nature.
7. Rahmani-Andebili, M. (2021). Precalculus – Practice Problems, Methods, and Solutions, Springer Nature.

Singularities of Complex Functions, Complex Series, Taylor and Laurent Series Expansions of Complex Functions, and Residue of Complex Functions: Solutions of Problems

6

Abstract

In this chapter, the problems of the fifth chapter are fully solved, in detail, step-by-step, and with different methods.

6.1 Poles, Removable Singularity, and Essential Singularity of a Complex Function

6.1. Based on the information given in the problem, we need to determine the poles of the complex function below [1–7]:

$$f(z) = \frac{z}{\sinh z \cosh z}$$

The roots of the denominator of a complex function are called the poles of the function.

Thus:

$$\sinh z \cosh z = 0$$

$$\Rightarrow 2 \sinh z \cosh z = 0 \Rightarrow \sinh 2z = 0$$

$$\Rightarrow 2z = k\pi i, k\epsilon\mathbb{Z} \Rightarrow z = \frac{k\pi}{2} i, k\epsilon\mathbb{Z}$$

However, $z = 0$ is not the pole of the function because both numerator and denominator of the function become zero in this condition. Herein, $z = 0$ is a removable singularity of the function, as can be seen in the following:

$$\lim_{z \to 0} \frac{z}{\sinh z \cosh z} = \lim_{z \to 0} \frac{2z}{\sinh 2z} \equiv \lim_{z \to 0} \frac{2z}{2z} = 1$$

Therefore:

$$z = i\frac{k\pi}{2}, k\epsilon\mathbb{Z} - \{0\}$$

Choice (1) is the answer.

© The Author(s), under exclusive license to Springer Nature Switzerland AG 2025
M. Rahmani-Andebili, *Mathematics of Engineering and Science*, https://doi.org/10.1007/978-3-031-71934-9_6

Notes

In this problem, the relations below have been used:

$$\sinh 2z = 2\sinh z \cosh z$$

$$\sinh z = 0 \Rightarrow z = k\pi i, k\epsilon \mathbb{Z}$$

$$\lim_{z\to 0} z\to 0 \sinh 2z \equiv \lim_{z\to 0} 2z$$

If the Laurent series expansion of a complex function includes an infinite number of terms with a negative exponent of $z - z_0$, the point $z = z_0$ for the function is an essential singular point. Moreover, if the Laurent series expansion includes a finite number of terms with a negative exponent of $z - z_0$, the point $z = z_0$ is called a pole. However, if the point $z = z_0$ is the root of both numerator and denominator, and the limit of the function at this point is finite, the point $z = z_0$ is called a removable singular point.

6.2. Based on the information given in the problem, we need to determine the type of the point $z = 0$ for the complex function below:

$$f(z) = \frac{1 - \cos z}{z^3}$$

The Laurent series expansion of the cosine term can be used in the main function as follows:

$$f(z) = \frac{1 - \left(1 - \frac{z^2}{2!} + \frac{z^4}{4!} - \cdots\right)}{z^3} = \frac{\frac{z^2}{2!} - \frac{z^4}{4!} + \cdots}{z^3} = \frac{1}{2!z} - \frac{1}{4!}z^2 + \cdots$$

As can be seen, the point $z = 0$ for the function is a first-order pole which is called a simple pole. Choice (2) is the answer.

Notes

In this problem, the relations below have been used:

$$\cos z = \sum_{n=0}^{\infty} (-1)^n \frac{z^{2n}}{(2n)!} = 1 - \frac{z^2}{2!} + \frac{z^4}{4!} - \cdots, \quad |z| < \infty$$

If the Laurent series expansion of a complex function includes an infinite number of terms with a negative exponent of $z - z_0$, the point $z = z_0$ for the function is an essential singular point. Moreover, if the Laurent series expansion includes a finite number of terms with a negative exponent of $z - z_0$, the point $z = z_0$ is called a pole. However, if the point $z = z_0$ is the root of both numerator and denominator, and the limit of the function at this point is finite, the point $z = z_0$ is called a removable singular point.

6.3. Based on the information given in the problem, we need to determine the order of the pole $z = 0$ of the complex function below:

$$f(z) = \frac{\cos z - 1}{z^4}$$

The order of the pole $z = z_0$ of a complex function $f(z)$ can be determined by finding the value of m in which the value of $\lim_{z\to z_0} (z - z_0)^m f(z)$ becomes finite for the first time.

Therefore:

$$\lim_{z \to 0} z^m \left(\frac{\cos z - 1}{z^4} \right) \equiv \lim_{z \to 0} z^m \frac{-\frac{z^2}{2}}{z^4} = \lim_{z \to 0} \left(\frac{-\frac{1}{2}}{z^{2-m}} \right)$$

As can be noticed, for $m = 2$, the value of the limit becomes finite for the first time. Thus, the pole $z = 0$ is a second-order pole for the complex function. Choice (2) is the answer.

Notes

In this problem, the relation below has been used:

$$\lim_{z \to 0} (1 - \cos z) \equiv \lim_{z \to 0} \frac{z^2}{2}$$

6.4. Based on the information given in the problem, we need to determine the number of singular points of the complex function below in the region $|z| < 2$:

$$f(z) = \frac{\ln(z + 3)}{(z^2 + 2) \sin z}$$

For the complex function $f(z) = \ln g(z)$, all the points on the line $Im\{g(z)\} = 0, \quad Re\{g(z)\} \le 0$ are nonholomorphic (nonanalytic).

Moreover, the number of singular points of the complex function $f(z) = \frac{p(z)}{q(z)}$ can be calculated by finding the number of roots of the equation $q(z) = 0$.

Therefore:

$$(z^2 + 2) \sin z = 0 \Rightarrow \begin{cases} \sin z = 0 \\ z^2 + 2 = 0 \end{cases}$$

$$\Rightarrow z = \pm i\sqrt{2}, 0, \pm \pi, \pm 2\pi, \cdots$$

As can be noticed, only $z = \pm i\sqrt{2}, 0$ are in the region $|z| < 2$.

On the other hand, $\ln(z + 3)$ is holomorphic (analytic) in the region $|z| < 2$.

Therefore, in total, three singular points of the complex function reside in the region $|z| < 2$. Choice (3) is the answer.

Notes

In this problem, the relation below has been used:

$$\sinh z = 0 \Rightarrow z = k\pi i, k \in \mathbb{Z}$$

6.5. Based on the information given in the problem, we need to determine the type of the point $z = 0$ for the complex function below:

$$f(z) = z^2 e^{\frac{1}{z}}$$

The Laurent series expansion of the exponential term can be used in the main function as follows:

$$z^2 e^{\frac{1}{z}} = z^2 \left(1 + \frac{1}{z} + \frac{1}{2!z^2} + \frac{1}{3!z^3} + \frac{1}{4!z^4} + \cdots\right) = z^2 + z + \frac{1}{2!} + \frac{1}{3!z} + \frac{1}{4!z^2} + \cdots$$

Since the Laurent series expansion of the complex function includes an infinite number of terms with a negative exponent of z, the point $z = 0$ for the function is an essential singular point. Choice (1) is the answer.

Notes

In this problem, the relations below have been used:

$$e^z = \sum_{n=0}^{\infty} \frac{z^n}{n!} = 1 + z + \frac{z^2}{2!} + \cdots$$

If the Laurent series expansion of a complex function includes an infinite number of terms with a negative exponent of $z - z_0$, the point $z = z_0$ for the function is an essential singular point. Moreover, if the Laurent series expansion includes a finite number of terms with a negative exponent of $z - z_0$, the point $z = z_0$ is called a pole. However, if the point $z = z_0$ is the root of both numerator and denominator, and the limit of the function at this point is finite, the point $z = z_0$ is called a removable singular point.

6.6. Based on the information given in the problem, we need to determine the type of the point $z = 0$ for the complex function below:

$$f(z) = \frac{e^z - 1}{z^2}$$

The order of the pole $z = z_0$ of a complex function $f(z)$ can be determined by finding the value of m in which the value of $\lim_{z \to z_0} (z - z_0)^m f(z)$ becomes finite for the first time.

Therefore:

$$\lim_{z \to 0} z^m \left(\frac{e^z - 1}{z^2}\right) = \lim_{z \to 0} \frac{e^z - 1}{z^{2-m}}$$

For $m = 0$, we have:

$$\lim_{z \to 0} \frac{e^z - 1}{z^2} = \frac{0}{0}$$

$$\overset{H}{\Rightarrow} \lim_{z \to 0} \frac{e^z}{2z} \to \infty$$

For $m = 1$, we have:

$$\lim_{z \to 0} \frac{e^z - 1}{z} = \frac{0}{0}$$

$$\overset{H}{\Rightarrow} \lim_{z \to 0} \frac{e^z}{1} = 1$$

As can be seen, for $m = 1$, the value of the limit becomes finite for the first time. Thus, the pole $z = 0$ is a first-order pole (simple pole) for the complex function. Choice (2) is the answer.

Notes

In this problem, the relation below has been used:

$$h(z) = \lim_{z \to z_0} \frac{f(z)}{g(z)} = \frac{0}{0} \overset{H}{\Rightarrow} h(z) = \lim_{z \to z_0} \frac{f'(z)}{g'(z)}$$

6.7. Based on the information given in the problem, we need to determine the type of the point $z = 1$ for the complex function below:

$$f(z) = e^{-\frac{1}{(z-1)^2}}$$

The Laurent series expansion of the function is as follows:

$$e^{-\frac{1}{(z-1)^2}} = 1 - \frac{1}{(z-1)^2} + \frac{1}{2!(z-1)^2} - \frac{1}{3!(z-1)^6} + \cdots$$

Since the Laurent series expansion of the complex function includes an infinite number of terms with a negative exponent of $(z - 1)$, the point $z = 1$ for the function is an essential singular point. Choice (2) is the answer.

Notes

In this problem, the relations below have been used:

$$e^z = \sum_{n=0}^{\infty} \frac{z^n}{n!} = 1 + z + \frac{z^2}{2!} + \cdots$$

If the Laurent series expansion of a complex function includes an infinite number of terms with a negative exponent of $z - z_0$, the point $z = z_0$ for the function is an essential singular point. Moreover, if the Laurent series expansion includes a finite number of terms with a negative exponent of $z - z_0$, the point $z = z_0$ is called a pole. However, if the point $z = z_0$ is the root of both numerator and denominator, and the limit of the function at this point is finite, the point $z = z_0$ is called a removable singular point.

6.8. Based on the information given in the problem, we need to determine the singular points of the following complex function:

$$f(z) = \cot(\pi z)$$

$$\Rightarrow f(z) = \frac{\cos(\pi z)}{\sin(\pi z)}$$

The singular points of the complex function $f(z) = \frac{p(z)}{q(z)}$ can be calculated by finding the roots of the equation $q(z) = 0$. Therefore:

$$\Rightarrow \sin(\pi z) = 0 \Rightarrow \pi z = 0, \pm \pi, \pm 2\pi, \cdots \Rightarrow z = 0, \pm 1, \pm 2, \cdots$$

Choice (3) is the answer.

Notes

In this problem, the relations below have been used:

$$\cot z = \frac{\cos z}{\sin z}$$

$$\sin z = 0 \Rightarrow z = k\pi, k\epsilon \mathbb{Z}$$

6.2   Complex Series

6.9. Based on the information given in the problem, the series is as follows:

$$\sum_{n=1}^{\infty} \left(1 - \frac{1}{n}\right)^{n^2} z^n \tag{1}$$

The general form of complex power series is as follows:

$$\sum_{n=1}^{\infty} C_n (z - z_0)^n \tag{2}$$

The series converges in the region $|z - z_0| < R$ in which R is the convergence radius of the series that can be calculated based on one of the following methods:

$$\frac{1}{R} = \lim_{n \to \infty} \sqrt[n]{|C_n|} \tag{3}$$

$$\frac{1}{R} = \lim_{n \to \infty} \left| \frac{C_{n+1}}{C_n} \right| \tag{4}$$

Solving (1) and (3):

$$\frac{1}{R} = \lim_{n \to \infty} \sqrt[n]{\left| \left(1 - \frac{1}{n}\right)^{n^2} \right|} = \lim_{n \to \infty} \left(1 - \frac{1}{n}\right)^{n} = e^{-1}$$

$$\Rightarrow R = e$$

Choice (2) is the answer.

Notes

In this problem, the relations below have been used:

$$\sqrt[m]{a^n} = a^{\frac{n}{m}}$$

$$\lim_{n \to \infty} n \to \infty \left(1 + \frac{a}{bn}\right)^{cn} = e^{\frac{ac}{b}}$$

6.10. Based on the information given in the problem, the series is as follows:

$$\sum_{n=0}^{\infty} \frac{(2n)!}{(n!)^2} z^n \tag{1}$$

The general form of complex power series is as follows:

$$\sum_{n=1}^{\infty} C_n (z - z_0)^n \tag{2}$$

The series converges in the region $|z - z_0| < R$ in which R is the convergence radius of the series that can be calculated based on one of the following methods:

$$\frac{1}{R} = \lim_{n \to \infty} \sqrt[n]{|C_n|} \tag{3}$$

$$\frac{1}{R} = \lim_{n \to \infty} \left| \frac{C_{n+1}}{C_n} \right| \tag{4}$$

Solving (1) and (4):

$$\frac{1}{R} = \lim_{n \to \infty} \left| \frac{\frac{(2(n+1))!}{((n+1)!)^2}}{\frac{(2n)!}{(n!)^2}} \right| = \lim_{n \to \infty} \frac{\frac{(2n+2)!}{((n+1)!)^2}}{\frac{(2n)!}{(n!)^2}}$$

$$\Rightarrow \frac{1}{R} = \lim_{n \to \infty} \frac{(2n+2)!(n!)^2}{((n+1)!)^2 (2n)!}$$

$$\Rightarrow \frac{1}{R} = \lim_{n \to \infty} \frac{(2n+2)(2n+1)(2n)!(n!)^2}{(n!)^2(n+1)^2(2n)!}$$

$$\Rightarrow \frac{1}{R} = \lim_{n \to \infty} \frac{(2n+2)(2n+1)}{(n+1)^2}$$

$$\Rightarrow \frac{1}{R} = \lim_{n \to \infty} \frac{4n^2+6n+2}{n^2+2n+1} \equiv \lim_{n \to \infty} \frac{4n^2}{n^2} = 4$$

$$\Rightarrow R = \frac{1}{4}$$

Choice (2) is the answer.

6.11. Based on the information given in the problem, the series is as follows:

$$\sum_{n=0}^{\infty} (\cos in)z^n = \sum_{n=0}^{\infty} (\cosh n)z^n \tag{1}$$

The general form of complex power series is as follows:

$$\sum_{n=1}^{\infty} C_n(z - z_0)^n \tag{2}$$

The series converges in the region $|z - z_0| < R$ in which R is the convergence radius of the series that can be calculated based on one of the following methods:

$$\frac{1}{R} = \lim_{n \to \infty} \sqrt[n]{|C_n|} \tag{3}$$

$$\frac{1}{R} = \lim_{n \to \infty} \left| \frac{C_{n+1}}{C_n} \right| \tag{4}$$

Solving (1) and (4):

$$\frac{1}{R} = \lim_{n \to \infty} \left| \frac{\cosh(n+1)}{\cosh n} \right|$$

$$\Rightarrow \frac{1}{R} = \lim_{n \to \infty} \left| \frac{\frac{e^{-(n+1)}+e^{(n+1)}}{2}}{\frac{e^{-n}+e^n}{2}} \right|$$

$$\Rightarrow \frac{1}{R} = \lim_{n \to \infty} \frac{e^{-(n+1)}+e^{(n+1)}}{e^{-n}+e^n} \equiv \lim_{n \to \infty} \frac{e^{n+1}}{e^n} = \lim_{n \to \infty} e = e$$

$$\Rightarrow R = e^{-1}$$

Choice (1) is the answer.

Notes

In this problem, the relations below have been used:

$$\cos in = \cosh n$$

$$\cosh n = \frac{e^{-n} + e^n}{2}$$

$$\frac{e^a}{e^b} = e^{a-b}$$

6.12. Based on the information given in the problem, we need to determine the convergence region of the complex series below:

$$f(z) = \sum_{n=1}^{\infty} e^{\frac{i}{z+1}n}$$

The criterion for the convergence of the series is as follows:

$$\left| e^{\frac{i}{z+1}} \right| < 1$$

As we know, $z = x + iy$. Therefore:

$$\left| e^{\frac{i}{x+iy+1}} \right| < 1 \Rightarrow \left| e^{\frac{i}{(x+1)+iy}} \right| < 1$$

$$\Rightarrow \left| e^{\frac{i}{(x+1)+iy} \times \frac{(x+1)-iy}{(x+1)-iy}} \right| < 1 \Rightarrow \left| e^{\frac{i(x+1)+y}{(x+1)^2+y^2}} \right| < 1$$

$$\Rightarrow \left| e^{\frac{i(x+1)}{(x+1)^2+y^2}} e^{\frac{y}{(x+1)^2+y^2}} \right| < 1 \Rightarrow \left| e^{\frac{i(x+1)}{(x+1)^2+y^2}} \right| \left| e^{\frac{y}{(x+1)^2+y^2}} \right| < 1$$

$$\Rightarrow e^{\frac{y}{(x+1)^2+y^2}} < 1 \Rightarrow \frac{y}{(x+1)^2 + y^2} < 0$$

$$\Rightarrow y < 0$$

Choice (2) is the answer.

Notes

In this problem, the relation below has been used:

$$\left| e^{ai} \right| = 1$$

6.3   Taylor and Laurent Series Expansions of Complex Functions

6.13. Based on the information given in the problem, we have:

$$f(z) = \ln\left(\frac{1+z}{1-z}\right) \ (1)$$

$$\Rightarrow f(z) = \ln(1+z) - \ln(1-z) \ (2)$$

The Taylor series expansion of a complex function can be calculated as follows:

$$f(z) = \sum_{n=0}^{\infty} C_n (z-z_0)^n = \sum_{n=0}^{\infty} \frac{1}{n!} f^{(n)}(z_0)(z-z_0)^n \ (3)$$

$$\Rightarrow f(z) = f(z_0) + f'(z_0)(z-z_0) + \frac{f''(z_0)}{2!}(z-z_0)^2 + \frac{f^{(3)}(z_0)}{3!}(z-z_0)^3 + \cdots (4)$$

where $f(z)$ is a complex holomorphic function in the given region and z_0 is in that region.

The Laurent series expansion of a complex function can be calculated by assuming $z_0 = 0$. In other words:

$$f(z) = \sum_{n=0}^{\infty} C_n z^n = \sum_{n=0}^{\infty} \frac{1}{n!} f^{(n)}(0) z^n = f(0) + f'(0)z + \frac{f''(0)}{2!} z^2 + \frac{f^{(3)}(0)}{3!} z^3 + \cdots (5)$$

Solving (2) and (5):

$$\Rightarrow f(z) = \left(z - \frac{z^2}{2} + \frac{z^3}{3} - \frac{z^4}{4} + \cdots\right) - \left(-z - \frac{z^2}{2} + \frac{z^3}{3} - \frac{z^4}{4} + \cdots\right) \ (6)$$

$$\Rightarrow f(z) = 2z + \frac{2z^3}{3} + \frac{2z^5}{5} + \cdots (7)$$

$$\Rightarrow f(z) = \sum_{n=0}^{\infty} \frac{2z^{2n+1}}{2n+1}$$

Choice (4) is the answer.

Notes

In this problem, the relation below has been used:

$$\ln \frac{a}{b} = \ln a - \ln b$$

6.14. Based on the information given in the problem, we have:

$$f(z) = \tan^{-1} z \quad (1)$$

As we know:

$$\tan^{-1} z = \int \frac{1}{1+z^2} dz \quad (2)$$

The problem can be solved using a heuristic method as follows:

As we know, the Laurent series expansion of $\frac{1}{1+z}$ is as follows:

$$\frac{1}{1+z} = 1 - z + z^2 - z^3 + \cdots \quad (3)$$

By replacing z by z^2 in (3), we can determine the Laurent series expansion of $\frac{1}{1+z^2}$ as follows:

$$\frac{1}{1+z^2} = 1 - z^2 + z^4 - z^6 + \cdots \quad (4)$$

Solving (1), (2), and (4):

$$f(z) = \int \left(1 - z^2 + z^4 - z^6 + \cdots \right) dz$$

$$f(z) = z - \frac{z^3}{3} + \frac{z^5}{5} - \cdots + c$$

By assuming $f(0) = 0$, we have $c = 0$. Hence:

$$f(z) = z - \frac{z^3}{3} + \frac{z^5}{5} - \cdots$$

Choice (4) is the answer.

Notes

In this problem, the relations below have been used:

The Taylor series expansion of a complex function can be calculated as follows:

$$f(z) = \sum_{n=0}^{\infty} C_n (z - z_0)^n = \sum_{n=0}^{\infty} \frac{1}{n!} f^{(n)}(z_0)(z - z_0)^n$$

$$\Rightarrow f(z) = f(z_0) + f'(z_0)(z - z_0) + \frac{f''(z_0)}{2!}(z - z_0)^2 + \frac{f^{(3)}(z_0)}{3!}(z - z_0)^3 + \cdots$$

where $f(z)$ is a complex holomorphic function in the given region and z_0 is in that region.

The Laurent series expansion of a complex function can be calculated by assuming $z_0 = 0$. In other words:

$$f(z) = \sum_{n=0}^{\infty} C_n z^n = \sum_{n=0}^{\infty} \frac{1}{n!} f^{(n)}(0) z^n = f(0) + f'(0)z + \frac{f''(0)}{2!} z^2 + \frac{f^{(3)}(0)}{3!} z^3 + \cdots$$

6.15. Based on the information given in the problem, we have:

$$f(z) = \frac{z}{(z-1)(z-3)} \tag{1}$$

$$1 < |z-1| < 2 \tag{2}$$

The function can be decomposed into two fractions as follows:

$$f(z) = \frac{z}{(z-1)(z-3)} = \frac{A}{z-1} + \frac{B}{z-3} \tag{3}$$

$$\Rightarrow \frac{z}{(z-1)(z-3)} = \frac{A(z-3) + B(z-1)}{(z-1)(z-3)} \tag{4}$$

$$\Rightarrow \frac{z}{(z-1)(z-3)} = \frac{(A+B)z - 3A - B}{(z-1)(z-3)} \tag{5}$$

$$\Rightarrow \begin{cases} A + B = 1 \\ -3A - B = 0 \end{cases} \Rightarrow A = -\frac{1}{2}, \quad B = \frac{3}{2} \tag{6}$$

$$\Rightarrow f(z) = -\frac{1}{2(z-1)} + \frac{3}{2(z-3)} \tag{7}$$

From the choices, it is noticed that the expansion needs to be around $z = 1$. Moreover, from (2), we have:

$$|z-1| < 2 \Rightarrow \left| \frac{z-1}{2} \right| < 1 \tag{8}$$

Therefore, the function needs to be written based on the positive exponents of the term $\frac{z-1}{2}$ as follows:

$$f(z) = -\frac{1}{2(z-1)} + \frac{3}{2((z-1) - 2)} \tag{9}$$

$$\Rightarrow f(z) = -\frac{1}{2(z-1)} - \frac{3}{4} \left(\frac{1}{1 - \frac{z-1}{2}} \right) \tag{10}$$

As we know, the Laurent series expansion of $\frac{1}{1-z}$ with the condition $|z| < 1$ is as follows:

$$\frac{1}{1-z} = 1 + z + z^2 + \cdots = \sum_{n=0}^{\infty} z^n \tag{11}$$

Solving (10) and (11):

$$f(z) = -\frac{1}{2(z-1)} - \frac{3}{4}\sum_{n=0}^{\infty}\left(\frac{z-1}{2}\right)^n$$

Choice (1) is the answer.

Notes

In this problem, the relations below have been used:

The Taylor series expansion of a complex function can be calculated as follows:

$$f(z) = \sum_{n=0}^{\infty} C_n(z-z_0)^n = \sum_{n=0}^{\infty}\frac{1}{n!}f^{(n)}(z_0)(z-z_0)^n$$

$$\Rightarrow f(z) = f(z_0) + f'(z_0)(z-z_0) + \frac{f''(z_0)}{2!}(z-z_0)^2 + \frac{f^{(3)}(z_0)}{3!}(z-z_0)^3 + \cdots$$

where $f(z)$ is a complex holomorphic function in the given region and z_0 is in that region.

The Laurent series expansion of a complex function can be calculated by assuming $z_0 = 0$. In other words:

$$f(z) = \sum_{n=0}^{\infty} C_n z^n = \sum_{n=0}^{\infty}\frac{1}{n!}f^{(n)}(0)z^n = f(0) + f'(0)z + \frac{f''(0)}{2!}z^2 + \frac{f^{(3)}(0)}{3!}z^3 + \cdots$$

6.16. Based on the information given in the problem, we have:

$$f(z) = \frac{1}{z-4} \tag{1}$$

$$|z| > 4 \tag{2}$$

$$\Rightarrow \left|\frac{4}{z}\right| < 1 \tag{3}$$

Therefore, the function needs to be written based on the positive exponents of the term $\frac{4}{z}$ as follows:

$$f(z) = \frac{1}{z}\frac{1}{1-\frac{4}{z}} \tag{4}$$

As we know, the Laurent series expansion of $\frac{1}{1-z}$ with the condition $|z| < 1$ is as follows:

$$\frac{1}{1-z} = 1 + z + z^2 + \cdots = \sum_{n=0}^{\infty} z^n \tag{5}$$

Solving (4) and (5):

$$f(z) = \frac{1}{z}\sum_{n=0}^{\infty}\left(\frac{4}{z}\right)^n = \sum_{n=0}^{\infty}\frac{4^n}{z^{n+1}}$$

$$\Rightarrow f(z) = \sum_{n=1}^{\infty}\frac{4^{n-1}}{z^n}$$

Choice (4) is the answer.

Notes

In this problem, the relations below have been used:

The Taylor series expansion of a complex function can be calculated as follows:

$$f(z) = \sum_{n=0}^{\infty} C_n (z-z_0)^n = \sum_{n=0}^{\infty}\frac{1}{n!}f^{(n)}(z_0)(z-z_0)^n$$

$$\Rightarrow f(z) = f(z_0) + f'(z_0)(z-z_0) + \frac{f''(z_0)}{2!}(z-z_0)^2 + \frac{f^{(3)}(z_0)}{3!}(z-z_0)^3 + \cdots$$

where $f(z)$ is a complex holomorphic function in the given region and z_0 is in that region.

The Laurent series expansion of a complex function can be calculated by assuming $z_0 = 0$. In other words:

$$f(z) = \sum_{n=0}^{\infty} C_n z^n = \sum_{n=0}^{\infty}\frac{1}{n!}f^{(n)}(0)z^n = f(0) + f'(0)z + \frac{f''(0)}{2!}z^2 + \frac{f^{(3)}(0)}{3!}z^3 + \cdots$$

6.4　　Residue of Complex Functions

6.17. Based on the information given in the problem, we need to calculate the residue of the following complex function at point $z_0 = 1$:

$$f(z) = \frac{z}{(z-1)(z+1)^2}$$

The residue of a function $f(z)$ at $z = z_0$ can be calculated by using two methods as follows:

Method 1: The coefficient of $\frac{1}{z-z_0}$ in the Laurent series expansion of a function $f(z)$ at $z = z_0$ is the residue.

Method 2: If z_0 is the mth-order pole of a complex function $f(z)$, then the residue of the function at $z = z_0$ can be calculated as follows:

$$\operatorname*{Res}_{z=z_0} f(z) = \frac{1}{(m-1)!} \lim_{z \to z_0} \frac{d^{m-1}}{dz^{m-1}} \left[(z - z_0)^m f(z) \right]$$

If z_0 is the first-order pole of a complex function $f(z)$, then the residue of the function at $z = z_0$ can be calculated as follows:

$$\operatorname*{Res}_{z=z_0} f(z) = \lim_{z \to z_0} (z - z_0) f(z)$$

Therefore:

$$\operatorname*{Res}_{z=1} f(z) = \lim_{z \to 1} (z - 1) f(z) = \lim_{z \to 1} \frac{z}{(z+1)^2}$$

$$\Rightarrow \operatorname*{Res}_{z=1} f(z) = \frac{1}{4}$$

Choice (2) is the answer.

6.18. Based on the information given in the problem, we need to calculate the residue of the following complex function at point $z_0 = -2$:

$$f(z) = \frac{1}{z(z+2)^3}$$

The residue of a function $f(z)$ at $z = z_0$ can be calculated by using two methods as follows:

Method 1: The coefficient of $\frac{1}{z - z_0}$ in the Laurent series expansion of a function $f(z)$ at $z = z_0$ is the residue.

Method 2: If z_0 is the mth-order pole of a complex function $f(z)$, then the residue of the function at $z = z_0$ can be calculated as follows:

$$\operatorname*{Res}_{z=z_0} f(z) = \frac{1}{(m-1)!} \lim_{z \to z_0} \frac{d^{m-1}}{dz^{m-1}} \left[(z - z_0)^m f(z) \right]$$

If z_0 is the first-order pole of a complex function $f(z)$, then the residue of the function at $z = z_0$ can be calculated as follows:

$$\operatorname*{Res}_{z=z_0} f(z) = \lim_{z \to z_0} (z - z_0) f(z)$$

Therefore:

$$\operatorname*{Res}_{z=-2} f(z) = \frac{1}{(3-1)!} \lim_{z \to -2} \frac{d^2}{dz^2} \left[(z+2)^3 \times \frac{1}{z(z+2)^3} \right]$$

$$\Rightarrow \operatorname*{Res}_{z=-2} f(z) = \frac{1}{2} \lim_{z \to -2} \left[\frac{1}{z} \right]'' = \frac{1}{2} \lim_{z \to -2} \left[-\frac{1}{z^2} \right]' = \frac{1}{2} \lim_{z \to -2} \frac{2}{z^3}$$

$$\Rightarrow \operatorname*{Res}_{z=-2} f(z) = -\frac{1}{8}$$

Choice (4) is the answer.

Notes

In this problem, the relation below has been used:

$$\left(\frac{u}{v}\right)' = \frac{u'v - v'u}{v^2}$$

6.19. Based on the information given in the problem, we need to calculate the residue of the following complex function at point $z_0 = -1$:

$$f(z) = \frac{e^z}{(z+1)^3 (z-2)}$$

The residue of a function $f(z)$ at $z = z_0$ can be calculated by using two methods as follows:

Method 1: The coefficient of $\frac{1}{z-z_0}$ in the Laurent series expansion of a function $f(z)$ at $z = z_0$ is the residue.

Method 2: If z_0 is the mth-order pole of a complex function $f(z)$, then the residue of the function at $z = z_0$ can be calculated as follows:

$$\operatorname*{Res}_{z=z_0} f(z) = \frac{1}{(m-1)!} \lim_{z \to z_0} \frac{d^{m-1}}{dz^{m-1}} [(z-z_0)^m f(z)]$$

If z_0 is the first-order pole of a complex function $f(z)$, then the residue of the function at $z = z_0$ can be calculated as follows:

$$\operatorname*{Res}_{z=z_0} f(z) = \lim_{z \to z_0} (z-z_0) f(z)$$

Therefore:

$$\operatorname*{Res}_{z=-1} f(z) = \frac{1}{(3-1)!} \lim_{z \to -1} \frac{d^2}{dz^2} \left[(z+1)^3 \times \frac{e^z}{(z+1)^3 (z-2)} \right]$$

$$\Rightarrow \operatorname*{Res}_{z=-1} f(z) = \frac{1}{2} \lim_{z \to -1} \left[\frac{e^z}{(z-2)} \right]''$$

$$\Rightarrow \operatorname*{Res}_{z=-1} f(z) = \frac{1}{2} \lim_{z \to -1} \left[\frac{e^z(z-2) - e^z}{(z-2)^2} \right]' = \frac{1}{2} \lim_{z \to -1} \left[\frac{e^z(z-3)}{(z-2)^2} \right]'$$

$$\Rightarrow \operatorname*{Res}_{z=-1} f(z) = \frac{1}{2} \lim_{z \to -1} \left[\frac{(e^z(z-3) + e^z)(z-2)^2 - 2(z-2)e^z(z-3)}{(z-2)^4} \right]$$

$$\Rightarrow \operatorname*{Res}_{z=-1} f(z) = \frac{1}{2} \lim_{z \to -1} \left[\frac{e^z \left((z-2)^3 - 2(z-2)(z-3) \right)}{(z-2)^4} \right]$$

$$\Rightarrow \operatorname*{Res}_{z=-1} f(z) = \frac{e^{-1} \left((-1-2)^3 - 2(-1-2)(-1-3) \right)}{2(-1-2)^4} = \frac{e^{-1}(27-24)}{2 \times 81}$$

$$\Rightarrow \operatorname*{Res}_{z=-1} f(z) = \frac{1}{54e}$$

Choice (4) is the answer.

Notes

In this problem, the relations below have been used:

$$\left(\frac{u}{v} \right)' = \frac{u'v - v'u}{v^2}$$

$$(e^z)' = e^z$$

$$(z^n)' = nz^{n-1}$$

6.20. Based on the information given in the problem, we need to calculate the residue of the complex function below at point $z = -2i$:

$$f(z) = \frac{z^2 - 2z}{(z+1)^2(z^2+4)}$$

The residue of a function $f(z)$ at $z = z_0$ can be calculated by using two methods as follows:

Method 1: The coefficient of $\frac{1}{z-z_0}$ in the Laurent series expansion of a function $f(z)$ at $z = z_0$ is the residue.

Method 2: If z_0 is the mth-order pole of a complex function $f(z)$, then the residue of the function at $z = z_0$ can be calculated as follows:

$$\operatorname*{Res}_{z=z_0} f(z) = \frac{1}{(m-1)!} \lim_{z \to z_0} \frac{d^{m-1}}{dz^{m-1}} \left[(z-z_0)^m f(z) \right]$$

If z_0 is the first-order pole of a complex function $f(z)$, then the residue of the function at $z = z_0$ can be calculated as follows:

$$\operatorname*{Res}_{z=z_0} f(z) = \lim_{z \to z_0} (z - z_0) f(z)$$

Therefore:

$$\operatorname*{Res}_{z=-2i} f(z) = \lim_{z \to -2i} \left((z + 2i) \frac{z^2 - 2z}{(z+1)^2(z^2+4)} \right)$$

$$\Rightarrow \operatorname*{Res}_{z=-2i} f(z) = \lim_{z \to -2i} \left(\frac{z^2 - 2z}{(z+1)^2(z-2i)} \right)$$

$$\Rightarrow \operatorname*{Res}_{z=-2i} f(z) = \frac{(-2i)^2 - 2(-2i)}{(-2i+1)^2(-2i-2i)}$$

$$\Rightarrow \operatorname*{Res}_{z=-2i} f(z) = \frac{4i^2 + 4i}{(-4+1-4i)(-4i)} = \frac{4i(1+i)}{(-3-4i)(-4i)}$$

$$\Rightarrow \operatorname*{Res}_{z=-2i} f(z) = \frac{1+i}{3+4i}$$

Choice (1) is the answer.

Notes

In this problem, the relations below have been used:

$$a^2 + b^2 = (a + ib)(a - ib)$$

$$i^2 = -1$$

6.21. Based on the information given in the problem, we need to calculate the residue of the complex function below at point $z = 0$:

$$f(z) = z^2 e^{\frac{1}{z}}$$

The residue of a function $f(z)$ at $z = z_0$ can be calculated by using two methods as follows:

Method 1: The coefficient of $\frac{1}{z-z_0}$ in the Laurent series expansion of a function $f(z)$ at $z = z_0$ is the residue.

Method 2: If z_0 is the mth-order pole of a complex function $f(z)$, then the residue of the function at $z = z_0$ can be calculated as follows:

$$\operatorname*{Res}_{z=z_0} f(z) = \frac{1}{(m-1)!} \lim_{z \to z_0} \frac{d^{m-1}}{dz^{m-1}} [(z - z_0)^m f(z)]$$

If z_0 is the first-order pole of a complex function $f(z)$, then the residue of the function at $z = z_0$ can be calculated as follows:

$$\operatorname*{Res}_{z=z_0} f(z) = \lim_{z \to z_0} (z - z_0) f(z)$$

The Laurent series expansion of the exponential term can be used in the main function as follows:

$$z^2 e^{\frac{1}{z}} = z^2 \left(1 + \frac{1}{z} + \frac{1}{2! z^2} + \frac{1}{3! z^3} + \frac{1}{4! z^4} + \cdots \right) = z^2 + z + \frac{1}{2!} + \frac{1}{3! z} + \frac{1}{4! z^2} + \cdots$$

Therefore, the residue of the function at point $z = 0$ is $\frac{1}{3!}$ or $\frac{1}{6}$. Choice (2) is the answer.

Notes

In this problem, the relation below has been used:

$$e^z = \sum_{n=0}^{\infty} \frac{z^n}{n!} = 1 + z + \frac{z^2}{2!} + \cdots$$

6.22. Based on the information given in the problem, we need to calculate the residue of the complex function below at point $z = 0$:

$$f(z) = \frac{1 - \cos z}{z^3}, \quad z \neq 0$$

The residue of a function $f(z)$ at $z = z_0$ can be calculated by using two methods as follows:

Method 1: The coefficient of $\frac{1}{z - z_0}$ in the Laurent series expansion of a function $f(z)$ at $z = z_0$ is the residue.

Method 2: If z_0 is the mth-order pole of a complex function $f(z)$, then the residue of the function at $z = z_0$ can be calculated as follows:

$$\operatorname*{Res}_{z=z_0} f(z) = \frac{1}{(m-1)!} \lim_{z \to z_0} \frac{d^{m-1}}{dz^{m-1}} [(z - z_0)^m f(z)]$$

If z_0 is the first-order pole of a complex function $f(z)$, then the residue of the function at $z = z_0$ can be calculated as follows:

$$\operatorname*{Res}_{z=z_0} f(z) = \lim_{z \to z_0} (z - z_0) f(z)$$

The Laurent series expansion of the cosine term can be used in the main function as follows:

$$f(z) = \frac{1 - \left(1 - \frac{z^2}{2!} + \frac{z^4}{4!} - \cdots \right)}{z^3} = \frac{\frac{z^2}{2!} - \frac{z^4}{4!} + \cdots}{z^3} = \frac{1}{2! z} - \frac{1}{4!} z^2 + \cdots$$

Therefore, the residue of the function at point $z = 0$ is $\frac{1}{2!}$ or $\frac{1}{2}$. Choice (1) is the answer.

Notes

In this problem, the relation below has been used:

$$\cos z = \sum_{n=0}^{\infty} (-1)^n \frac{z^{2n}}{(2n)!} = 1 - \frac{z^2}{2!} + \frac{z^4}{4!} - \cdots, \quad |z| < \infty$$

6.23. Based on the information given in the problem, we need to calculate the residue of the following complex function at point $z_0 = 0$:

$$f(z) = \frac{\sinh z}{z^4}$$

The residue of a function $f(z)$ at $z = z_0$ can be calculated by using two methods as follows:

Method 1: The coefficient of $\frac{1}{z - z_0}$ in the Laurent series expansion of a function $f(z)$ at $z = z_0$ is the residue.

Method 2: If z_0 is the mth-order pole of a complex function $f(z)$, then the residue of the function at $z = z_0$ can be calculated as follows:

$$\operatorname*{Res}_{z=z_0} f(z) = \frac{1}{(m-1)!} \lim_{z \to z_0} \frac{d^{m-1}}{dz^{m-1}} [(z - z_0)^m f(z)]$$

If z_0 is the first-order pole of a complex function $f(z)$, then the residue of the function at $z = z_0$ can be calculated as follows:

$$\operatorname*{Res}_{z=z_0} f(z) = \lim_{z \to z_0} (z - z_0) f(z)$$

As we know, the Laurent series expansion of $\sinh z$ is as follows:

$$\sinh z = \sum_{n=0}^{\infty} \frac{z^{2n+1}}{(2n+1)!} = \frac{z}{1!} + \frac{z^3}{3!} + \frac{z^5}{5!} + \cdots, \quad |z| < \infty$$

Therefore:

$$\frac{\sinh z}{z^4} = \frac{1}{z^4}\left(z + \frac{z^3}{3!} + \frac{z^5}{5!} + \frac{z^7}{7!} + \cdots\right) = \frac{1}{z^3} + \frac{1}{3!}\frac{1}{z} + \frac{z}{5!} + \frac{z^2}{7!} + \cdots$$

Thus, the residue of the function at point $z_0 = 0$ is $\frac{1}{3!}$ or $\frac{1}{6}$. Choice (2) is the answer.

6.24. Based on the information given in the problem, we need to calculate the residue of the following complex function at point $z_0 = 0$:

$$f(z) = z^7 \cos \frac{1}{z^2}$$

The residue of a function $f(z)$ at $z = z_0$ can be calculated by using two methods as follows:

Method 1: The coefficient of $\frac{1}{z-z_0}$ in the Laurent series expansion of a function $f(z)$ at $z = z_0$ is the residue.

Method 2: If z_0 is the mth-order pole of a complex function $f(z)$, then the residue of the function at $z = z_0$ can be calculated as follows:

$$\operatorname*{Res}_{z=z_0} f(z) = \frac{1}{(m-1)!} \lim_{z \to z_0} \frac{d^{m-1}}{dz^{m-1}} [(z - z_0)^m f(z)]$$

If z_0 is the first-order pole of a complex function $f(z)$, then the residue of the function at $z = z_0$ can be calculated as follows:

$$\operatorname*{Res}_{z=z_0} f(z) = \lim_{z \to z_0} (z - z_0) f(z)$$

As we know, the Laurent series expansion of $\cos z$ is as follows:

$$\cos z = \sum_{n=0}^{\infty} (-1)^n \frac{z^{2n}}{(2n)!} = 1 - \frac{z^2}{2!} + \frac{z^4}{4!} - \frac{z^6}{6!} + \cdots, \quad |z| < \infty$$

Therefore:

$$\cos\left(\frac{1}{z^2}\right) = 1 - \frac{\left(\frac{1}{z^2}\right)^2}{2!} + \frac{\left(\frac{1}{z^2}\right)^2}{4!} - \frac{\left(\frac{1}{z^2}\right)^6}{6!} + \cdots$$

$$\Rightarrow z^7 \cos\left(\frac{1}{z^2}\right) = z^7 - \frac{z^3}{2!} + \frac{1}{4!z} - \frac{1}{6!z^5} + \cdots$$

Hence, the residue of the function at point $z_0 = 0$ is $\frac{1}{4!}$ or $\frac{1}{24}$. Choice (3) is the answer.

6.25. Based on the information given in the problem, we need to calculate the residue of the complex function below at point $z = \frac{3\pi}{2}$:

$$f(z) = e^z \tan z$$

The residue of a function $f(z)$ at $z = z_0$ can be calculated by using two methods as follows:

Method 1: The coefficient of $\frac{1}{z-z_0}$ in the Laurent series expansion of a function $f(z)$ at $z = z_0$ is the residue.

Method 2: If z_0 is the mth-order pole of a complex function $f(z)$, then the residue of the function at $z = z_0$ can be calculated as follows:

$$\operatorname*{Res}_{z=z_0} f(z) = \frac{1}{(m-1)!} \lim_{z \to z_0} \frac{d^{m-1}}{dz^{m-1}} [(z - z_0)^m f(z)]$$

If z_0 is the first-order pole of a complex function $f(z)$, then the residue of the function at $z = z_0$ can be calculated as follows:

$$\operatorname*{Res}_{z=z_0} f(z) = \lim_{z \to z_0} (z - z_0) f(z)$$

Therefore:

$$\operatorname{Res}_{z=\frac{3\pi}{2}} f(z) = \lim_{z \to \frac{3\pi}{2}} \left(z - \frac{3\pi}{2}\right) e^z \tan z = \lim_{z \to \frac{3\pi}{2}} \left(z - \frac{3\pi}{2}\right) e^z \frac{\sin z}{\cos z}$$

$$\Rightarrow \operatorname{Res}_{z=\frac{3\pi}{2}} f(z) = \left(\lim_{z \to \frac{3\pi}{2}} \frac{z - \frac{3\pi}{2}}{\cos z} \right) \times \left(\lim_{z \to \frac{3\pi}{2}} e^z \sin z \right)$$

$$\Rightarrow \operatorname{Res}_{z=\frac{3\pi}{2}} f(z) = \left(\frac{0}{0}\right) \times \left(\lim_{z \to \frac{3\pi}{2}} e^z \sin z \right)$$

$$\stackrel{H}{\Rightarrow} \operatorname{Res}_{z=\frac{3\pi}{2}} f(z) = \left(\lim_{z \to \frac{3\pi}{2}} \frac{1}{-\sin z} \right) \times \left(\lim_{z \to \frac{3\pi}{2}} e^z \sin z \right)$$

$$\Rightarrow \operatorname{Res}_{z=\frac{3\pi}{2}} f(z) = \left(\frac{1}{-\sin \frac{3\pi}{2}} \right) \times \left(e^{\frac{3\pi}{2}} \sin \frac{3\pi}{2} \right)$$

$$\Rightarrow \operatorname{Res}_{z=\frac{3\pi}{2}} f(z) = -e^{\frac{3\pi}{2}}$$

Choice (3) is the answer.

Notes

In this problem, the relations below have been used:

$$\tan z = \frac{\sin z}{\cos z}$$

$$\lim_{z \to z_0} (f(z)g(z)) = \left(\lim_{z \to z_0} f(z) \right) \times \left(\lim_{z \to z_0} g(z) \right)$$

$$h(z) = \lim_{z \to z_0} \frac{f(z)}{g(z)} = \frac{0}{0} \stackrel{H}{\Rightarrow} h(z) = \lim_{z \to z_0} \frac{f'(z)}{g'(z)}$$

$$\frac{d}{dz} (\cos z) = -\sin z$$

$$\sin \frac{3\pi}{2} = -1$$

References

1. Rahmani-Andebili, M. (2024). Precalculus (2nd Ed.) – Practice Problems, Methods, and Solutions, Springer Nature.
2. Rahmani-Andebili, M. (2023). Calculus III – Practice Problems, Methods, and Solutions, Springer Nature.
3. Rahmani-Andebili, M. (2023). Calculus II – Practice Problems, Methods, and Solutions, Springer Nature.
4. Rahmani-Andebili, M. (2023). Calculus I (2nd Ed.) – Practice Problems, Methods, and Solutions, Springer Nature.
5. Rahmani-Andebili, M. (2022). Differential Equations – Practice Problems, Methods, and Solutions, Springer Nature.
6. Rahmani-Andebili, M. (2021). Calculus – Practice Problems, Methods, and Solutions, Springer Nature.
7. Rahmani-Andebili, M. (2021). Precalculus – Practice Problems, Methods, and Solutions, Springer Nature.

Abstract

In this chapter, the basic and advanced problems of complex integration are presented. The subjects include complex integration of nonholomorphic functions, complex integration of holomorphic functions, and complex integration of functions including a finite number of singular points. Herein, different types of problems and exercises are presented that are categorized as follows:

o **Problems with detailed solution**: They have been designed to teach students the subjects in detail. Moreover, they have been categorized into different levels based on their difficulty levels (easy, normal, and hard) and calculation amounts (small, normal, and large).

o **Partially solved exercises**: They have been designed to encourage students to practice more problems while guiding them through the problem-solving procedure and hinting the required formulas.

o **Exercises with final answer**: They have been designed to encourage students to practice by themselves while hinting them by the final answer as well as to help instructors to give tests or quizzes.

7.1 Complex Integration of Nonholomorphic Functions

7.1. Calculate the following integral on the given curve for $0 \leq t \leq 2$ [1–7]:

$$I = \int_C \bar{z}\, dz, \quad C : \begin{cases} x = t^2 \\ y = t \end{cases}$$

Difficulty level   o Easy   • Normal   o Hard
Calculation amount   o Small   • Normal   o Large

1) $8 + \frac{16}{3}i$

2) $8 - \frac{16}{3}i$

3) $10 + \frac{8}{3}i$

4) $10 - \frac{8}{3}i$

Partially Solved Exercise

Calculate the following integral if $f(z) = Re\ (z)$ and C is the parabola $y = x^2$ form $z = 0$ to $z = 1 + i$:

$$I = \int_C f(z)dz$$

Solution

Based on the information given in the problem, we have:

$$I = \int_C f(z)dz \tag{1}$$

$$f(z) = Re(z) \tag{2}$$

$$C : y = x^2 \tag{3}$$

$$0 \le x, y \le 1 \tag{4}$$

As we know, $z = x + iy$. Hence, from (3), we have:

$$z = x + ix^2 \tag{5}$$

$$\Rightarrow dz = (\qquad)dx \tag{6}$$

Solving (1), (2), and (6):

$$I = \int_0^1 (\qquad)dx$$

$$\Rightarrow I = [\qquad]_0^1$$

$$\Rightarrow I = \frac{1}{2} + \frac{2}{3}i$$

Notes

In this problem, the relations below have been used:

$$Re(z) = x$$

$$\int x^n dx = \frac{1}{n+1}x^{n+1} + c$$

7.2. Calculate the following integral if C is a circle with radius 1 and the center located at $1 + i$:

$$I = \oint_C z d\bar{z}$$

Difficulty level   ○ Easy   ● Normal   ○ Hard
Calculation amount   ○ Small ● Normal ○ Large
1) 0
2) $-2\pi i$
3) $(1 - i)2\pi$
4) $2\pi i$

Partially Solved Exercise

Calculate the following integral:

$$I = \oint_{C:|z|=1} \bar{z} dz$$

Solution

The equation of the contour in polar coordinates is as follows:

$$z = e^{i\theta}, \quad 0 \leq \theta < 2\pi$$

Hence:

$$\bar{z} = e^{-i\theta}$$

$$dz = ie^{i\theta} d\theta$$

Therefore:

$$I = \int_0^{2\pi} (\qquad) d\theta$$

$$\Rightarrow I = \int_0^{2\pi} (\qquad) d\theta$$

$$\Rightarrow I = [\qquad]_0^{2\pi}$$

$$\Rightarrow I = 2\pi i$$

Notes

In this problem, the relations below have been used:

$$\int x^n dx = \frac{1}{n+1} x^{n+1} + c$$

The equation of a circle in Cartesian and polar coordinates with the radius r and the center at z_0 are as follows, respectively:

$$|z - z_0| = r$$

$$z = z_0 + re^{i\theta}, \quad 0 \le \theta < 2\pi$$

7.2 Complex Integration of Holomorphic Functions

7.3. Calculate the integral below if the contour C is a counterclockwise circle with the equation $|z + 1 - 2i| = 1$.

$$I = \oint_C \frac{z+i}{z(z+2)}$$

Difficulty level   ○ Easy   ● Normal   ○ Hard
Calculation amount   ○ Small   ● Normal   ○ Large
1) 0
2) πi
3) $-\frac{3}{2}\pi i$
4) $\frac{3}{2}\pi i$

Partially Solved Exercise

Calculate the integral below if the contour C is a counterclockwise circle with the equation $|z - 1 - 2i| = 2$.

$$I = \oint_C \frac{z+1}{z^3 - 2z^2}$$

Solution

The singular points of the complex function can be calculated as follows:

$$z^3 - 2z^2 = 0 \Rightarrow z =$$

Now, we need to see if these points are inside the circle or not.

$$z = \qquad \Rightarrow$$

$$z = \qquad \Rightarrow$$

$$\Rightarrow I = 0$$

Notes

In this problem, the relations below have been used:

$$|a + ib| = \sqrt{a^2 + b^2}$$

The singular points of the complex function $f(z) = \frac{p(z)}{q(z)}$ can be determined by finding the roots of the equation $q(z) = 0$.

For a circle with equation $|z - z_0| = R$, the point z_1 is inside the circle, on the circle, and outside the circle, respectively, if:

$$|z_1 - z_0| < R$$

$$|z_1 - z_0| = R$$

$$|z_1 - z_0| > R$$

Cauchy's integral theorem states that the integral around the contour C is zero if $f(z)$ on and inside the loop is holomorphic everywhere.

$$I = \oint_C f(z)dz = 0$$

Exercise

Calculate the integral below. Herein, the direction of contour is counterclockwise.

$$I = \oint_{|z|=1} z^3 \sin z \, dz$$

Final Answer

0

7.4. Calculate the following integral if $f(z) = z^3 \bar{z} \cos z$. Herein, the direction of contour is counterclockwise:

$$I = \oint_{C:|z|=1} f(z)dz$$

Difficulty level   ○ Easy   ○ Normal   ● Hard

Calculation amount   ● Small   ○ Normal   ○ Large

1) 0
2) πi
3) $-2\pi i$
4) 2π

7.3   Complex Integration of Functions Including Finite Number of Singular Points

7.5. Calculate the integral below if the contour C is a counterclockwise simple closed loop that surrounds the origin.

$$I = \oint_C \frac{dz}{z}$$

Difficulty level   ● Easy   ○ Normal   ○ Hard
Calculation amount   ● Small   ○ Normal   ○ Large
1) 0
2) $2\pi i$
3) $3\pi i$
4) πi

Exercise

Calculate the integral below if the contour C is a clockwise simple closed loop that surrounds the origin.

$$I = \oint_C \frac{2dz}{z}$$

Final Answer
$-4\pi i$

7.6. Calculate the integral below in which the contour C is a clockwise simple closed loop that surrounds $z = 1$.

$$I = \oint_C \frac{\cos \pi z}{z - 1}$$

Difficulty level   ● Easy   ○ Normal   ○ Hard
Calculation amount   ● Small   ○ Normal   ○ Large
1) $-2\pi i$
2) $2\pi i$
3) $4\pi i$
4) $-4\pi i$

7.7. Calculate the integral below. Herein, the direction of contour is clockwise.

$$I = \oint_{|z|=2} \frac{\cosh iz}{z^2 + 4z + 3}$$

Difficulty level   ○ Easy   ● Normal   ○ Hard
Calculation amount   ○ Small   ● Normal   ○ Large

1) πi
2) $-\pi i \cos 1$
3) $-\pi \cos 1$
4) $i \cos 1$

Exercise

Calculate the following integral: Herein, the direction of contour is clockwise.

$$I = \oint_C \frac{dz}{z-3}, C : |z+i| = 4$$

Final Answer

$-2\pi i$

7.8. Calculate the integral below. Herein, the direction of contour is counterclockwise.

$$I = \oint_C \frac{z+1}{z^3 - 2z^2} dz, \quad C : |z - 2 - i| = 2$$

Difficulty level   ○ Easy   ● Normal   ○ Hard
Calculation amount   ○ Small   ● Normal   ○ Large

1) 0
2) $3\pi i$
3) $-\dfrac{3\pi i}{2}$
4) $\dfrac{3\pi i}{2}$

7.9. Calculate the integral below if the contour C is a counterclockwise square with the corners located at ± 4 and $\pm 4i$.

$$I = \oint_C \frac{e^z}{z^4} dz$$

Difficulty level   ○ Easy   ● Normal   ○ Hard
Calculation amount   ○ Small   ● Normal   ○ Large

1) $-\dfrac{\pi i}{3}$
2) $\dfrac{\pi i}{3}$
3) $\dfrac{4\pi i}{3!}$
4) $-\dfrac{4\pi i}{3!}$

Exercise

Calculate the following integral: Herein, the direction of contour is clockwise.

$$I = \oint_C \frac{2dz}{z-2}, \quad C : |z| = 1$$

Final Answer

0

Partially Solved Exercise

Calculate the integral below if the contour C is a counterclockwise circle with the equation $|z| = 2$.

$$I = \oint_C \frac{e^z}{(z+1)^4} dz$$

Solution

If $f(z)$ is holomorphic on and inside the counterclockwise contour C (a closed loop that does not intersect itself) except at the finite number of singular points $z_1, z_2, . z_n$ and the contour surrounds these singular points, then the contour integral can be calculated as follows:

$$I = \oint_C f(z)dz = 2\pi i \left(\operatorname*{Res}_{z=z_1} f(z) + \operatorname*{Res}_{z=z_2} f(z) + \ldots + \operatorname*{Res}_{z=z_n} f(z) \right)$$

Moreover, if the direction of the contour C is clockwise, the contour integral is calculated as follows:

$$I = \oint_C f(z)dz = -2\pi i \left(\operatorname*{Res}_{z=z_1} f(z) + \operatorname*{Res}_{z=z_2} f(z) + \ldots + \operatorname*{Res}_{z=z_n} f(z) \right)$$

The singular points of the complex function can be calculated as follows:

$$(z+1)^4 = 0 \Rightarrow z = -1$$

Now, we need to see if the singular point is inside the circle or not.

$$z = -1 \Rightarrow |-1| = 1 < 2$$

As can be seen, the singular point, which is a four-order pole, is inside the contour. Therefore:

$$I = 2\pi i \left[\operatorname*{Res}_{z=-1} f(z) \right]$$

$$\Rightarrow I = 2\pi i \left[\frac{1}{3!} \lim_{z \to -1} \frac{d^3}{dz^3} [(\qquad) \times (\qquad)] \right]$$

$$\Rightarrow I = (\qquad) \left[\lim_{z \to -1} \frac{d^3}{dz^3} [\qquad] \right]$$

$$\Rightarrow I = (\qquad) [\lim_{z \to -1} [\qquad]]$$

$$\Rightarrow I = \frac{\pi i}{3e}$$

Notes

In this problem, the relations below have been used:

$$\frac{d}{dz} e^z = e^z$$

The singular points of the complex function $f(z) = \frac{p(z)}{q(z)}$ can be determined by finding the roots of the equation $q(z) = 0$.

For a circle with the equation $|z - z_0| = R$, the point z_1 is inside the circle, on the circle, and outside the circle, respectively, if:

$$|z_1 - z_0| < R$$

$$|z_1 - z_0| = R$$

$$|z_1 - z_0| > R$$

The residue of the function $f(z)$ at the given singularity can be calculated by using two methods as follows:

Method 1: The coefficient of $\frac{1}{z - z_0}$ in the Taylor series expansion of the function $f(z)$ at the essential singular point $z = z_0$ as well as the coefficient of $\frac{1}{z}$ in the Laurent series expansion of the function $f(z)$ at the essential singular point $z = 0$ is called residue.

Method 2: If z_0 is the mth-order pole of a complex function $f(z)$, then the residue of the function at $z = z_0$ can be calculated as follows:

$$\operatorname*{Res}_{z=z_0} f(z) = \frac{1}{(m-1)!} \lim_{z \to z_0} \frac{d^{m-1}}{dz^{m-1}} [(z - z_0)^m f(z)]$$

If z_0 is the first-order pole of a complex function $f(z)$, then the residue of the function at $z = z_0$ can be calculated as follows:

$$\operatorname*{Res}_{z=z_0} f(z) = \lim_{z \to z_0} (z - z_0) f(z)$$

7.10. Calculate the following integral: Herein, the direction of contour is clockwise.

$$I = \oint_{|z-1|=1} \frac{\sin \pi z}{(z^2 - 1)^2}$$

Difficulty level   ○ Easy   ○ Normal   ● Hard
Calculation amount   ○ Small   ● Normal   ○ Large

1) $\dfrac{\pi^2}{2}$

2) $-\dfrac{\pi^2}{2}$

3) $-\dfrac{\pi^2}{2}i$

4) $\dfrac{\pi^2}{2}i$

Partially Solved Exercise

Calculate the integral below if the contour C is a clockwise circle with the equation $|z| = 3$.

$$I = \oint_C \frac{z^3 - i}{\pi z}$$

Solution

If $f(z)$ is holomorphic on and inside the counterclockwise contour C (a closed loop that does not intersect itself) except at the finite number of singular points $z_1, z_2, , z_n$ and the contour surrounds these singular points, then the contour integral can be calculated as follows:

$$I = \oint_C f(z)dz = 2\pi i \left(\operatorname*{Res}_{z=z_1} f(z) + \operatorname*{Res}_{z=z_2} f(z) + \ldots + \operatorname*{Res}_{z=z_n} f(z) \right)$$

Moreover, if the direction of the contour C is clockwise, the contour integral is calculated as follows:

$$I = \oint_C f(z)dz = -2\pi i \left(\operatorname*{Res}_{z=z_1} f(z) + \operatorname*{Res}_{z=z_2} f(z) + \ldots + \operatorname*{Res}_{z=z_n} f(z) \right)$$

Herein, $z = 0$ is the only singular point which is inside the contour $|z| = 3$, as can be seen in the following:

$$z = 0 \Rightarrow |0| = 0 < 3$$

Therefore:

$$I = -2\pi i \left[\operatorname*{Res}_{z=0} f(z) \right]$$

$$\Rightarrow I = -2\pi i \left[\lim_{z \to 0} (\quad)(\quad) \right]$$

$$\Rightarrow I = -2\pi i \left[\lim_{z \to 0} (\quad) \right]$$

$$\Rightarrow I = -2\pi i \times (\quad)$$

$$\Rightarrow I = -2$$

Notes

In this problem, the relations below have been used:

$$i^2 = -1$$

The singular points of the complex function $f(z) = \frac{p(z)}{q(z)}$ can be determined by finding the roots of the equation $q(z) = 0$.

For a circle with equation $|z - z_0| = R$, the point z_1 is inside the circle, on the circle, and outside the circle, respectively, if:

$$|z_1 - z_0| < R$$

$$|z_1 - z_0| = R$$

$$|z_1 - z_0| > R$$

The residue of the function $f(z)$ at the given singularity can be calculated by using two methods as follows:

Method 1: The coefficient of $\frac{1}{z - z_0}$ in the Taylor series expansion of the function $f(z)$ at the essential singular point $z = z_0$ as well as the coefficient of $\frac{1}{z}$ in the Laurent series expansion of the function $f(z)$ at the essential singular point $z = 0$ is called residue.

Method 2: If z_0 is the *m*th-order pole of a complex function $f(z)$, then the residue of the function at $z = z_0$ can be calculated as follows:

$$\operatorname*{Res}_{z=z_0} f(z) = \frac{1}{(m-1)!} \lim_{z \to z_0} \frac{d^{m-1}}{dz^{m-1}} [(z - z_0)^m f(z)]$$

If z_0 is the first-order pole of a complex function $f(z)$, then the residue of the function at $z = z_0$ can be calculated as follows:

$$\operatorname*{Res}_{z=z_0} f(z) = \lim_{z \to z_0} (z - z_0) f(z)$$

Exercise

Calculate the integral below. Herein, the direction of contour is counterclockwise.

$$I = \oint_{|z|=2} \frac{e^z}{z - i} dz$$

Final Answer

$2\pi i e^i$

7.11. Calculate the integral below. Herein, the direction of contour is counterclockwise.

$$I = \oint_{|z|=1} \left(z + \frac{1}{z}\right) e^{\frac{1}{z}} dz$$

Difficulty level   ○ Easy   ● Normal   ○ Hard
Calculation amount   ○ Small   ● Normal   ○ Large
1) 0
2) πi
3) $2\pi i$
4) $3\pi i$

Exercise

Calculate the following integral: Herein, the direction of contour is counterclockwise.

$$I = \oint_{C:|z|=1} \frac{e^{iz}}{\sin 2z}$$

Final Answer

πi

7.12. Calculate the integral below if the contour C is a counterclockwise square with the corners located at $\pm\sqrt{2}$ and $\pm i\sqrt{2}$.

$$I = \oint_C \frac{\cosh \pi z}{z(z^2 + 1)}$$

Difficulty level   ○ Easy   ○ Normal   ● Hard
Calculation amount   ○ Small   ● Normal   ○ Large
1) $-4\pi i$
2) $-2\pi i$
3) $4\pi i$
4) $2\pi i$

Exercise

Calculate the integral below. Herein, the direction of contour is counterclockwise.

$$I = \oint_{|z|=4} \frac{z+1}{z^2+9} dz$$

Final Answer

$2\pi i$

7.13. Calculate the integral below. Herein, the direction of contour is counterclockwise.

$$I = \oint_{|z|=2} \tan z \, dz$$

Difficulty level   ○ Easy   ○ Normal   ● Hard
Calculation amount   ○ Small   ○ Normal   ● Large
1) πi
2) $4\pi i$
3) $-4\pi i$
4) $-\pi i$

Exercise

Calculate the integral below in which $C : |z| = 3$. Herein, the direction of contour is counterclockwise.

$$I = \oint_C \frac{\sin(\pi z^2) + \cos(\pi z^2)}{(z-1)(z-2)} \, dz$$

Final Answer
$4\pi i$

7.14. Calculate the following integral if C is $|z| = \frac{1}{3}$: Herein, the direction of contour is counterclockwise.

$$I = \oint_C 120 z^3 e^{-\frac{1}{z}} dz$$

Difficulty level   ○ Easy   ● Normal   ○ Hard
Calculation amount   ● Small   ○ Normal   ○ Large
1) $120\pi i$
2) $-240\pi i$
3) $-40\pi i$
4) $10\pi i$

Partially Solved Exercise

Calculate the following integral in counterclockwise direction:

$$I = \oint_{C:|z|=2} z e^{\frac{1}{z}} dz$$

Solution

If $f(z)$ is holomorphic on and inside the counterclockwise contour C (a closed loop that does not intersect itself) except at the finite number of singular points $z_1, z_2, , z_n$ and the contour surrounds these singular points, then the contour integral can be calculated as follows:

$$I = \oint_C f(z)dz = 2\pi i \left(\operatorname*{Res}_{z=z_1} f(z) + \operatorname*{Res}_{z=z_2} f(z) + \ldots + \operatorname*{Res}_{z=z_n} f(z) \right)$$

Moreover, if the direction of the contour C is clockwise, the contour integral is calculated as follows:

$$I = \oint_C f(z)dz = -2\pi i \left(\operatorname*{Res}_{z=z_1} f(z) + \operatorname*{Res}_{z=z_2} f(z) + \ldots + \operatorname*{Res}_{z=z_n} f(z) \right)$$

By replacing $e^{\frac{1}{z}}$ by its Laurent series expansions, we have:

$$ze^{\frac{1}{z}} = z(\qquad\qquad\qquad\qquad)$$

As can be seen, $z = 0$ is the essential singular point of the function because its Laurent series expansion includes an infinite number of terms with a negative exponent of z. Moreover, the singular point resides inside the contour, as can be seen in the following:

$$z = 0 \Rightarrow |0| = 0 < 2$$

The coefficient of $\frac{1}{z}$ as the residue of the function at $z = 0$ is as follows:

$$\operatorname*{Res}_{z=0} f(z) = (\qquad)$$

Therefore:

$$I = 2\pi i \times (\qquad)$$

$$\Rightarrow I = \pi i$$

Notes

In this problem, the relations below have been used:

The Laurent series expansion of e^z is as follows:

$$e^z = \sum_{n=0}^{\infty} \frac{z^n}{n!} = 1 + z + \frac{z^2}{2!} + \cdots, |z| < \infty$$

For a circle with equation $|z - z_0| = R$, the point z_1 is inside the circle, on the circle, and outside the circle, respectively, if:

$$|z_1 - z_0| < R$$

$$|z_1 - z_0| = R$$

$$|z_1 - z_0| > R$$

The residue of the function $f(z)$ at the given singularity can be calculated by using two methods as follows:

Method 1: The coefficient of $\frac{1}{z-z_0}$ in the Taylor series expansion of the function $f(z)$ at the essential singular point $z = z_0$ as well as the coefficient of $\frac{1}{z}$ in the Laurent series expansion of the function $f(z)$ at the essential singular point $z = 0$ is called residue.

Method 2: If z_0 is the mth-order pole of a complex function $f(z)$, then the residue of the function at $z = z_0$ can be calculated as follows:

$$\operatorname*{Res}_{z=z_0} f(z) = \frac{1}{(m-1)!} \lim_{z \to z_0} \frac{d^{m-1}}{dz^{m-1}} [(z-z_0)^m f(z)]$$

If z_0 is the first-order pole of a complex function $f(z)$, then the residue of the function at $z = z_0$ can be calculated as follows:

$$\operatorname*{Res}_{z=z_0} f(z) = \lim_{z \to z_0} (z - z_0) f(z)$$

Exercise

Calculate the integral below. Herein, the direction of contour is counterclockwise.

$$I = \oint_{|z|=1} z^2 e^{\frac{4}{z^3}} dz$$

Final Answer

$8\pi i$

7.15. Calculate the following integral on the unit circle centered at the origin with counterclockwise direction:

$$I = \oint_C e^{-\frac{1}{z}} \sin\left(\frac{2}{z}\right) dz$$

Difficulty level   ○ Easy   ● Normal   ○ Hard
Calculation amount   ● Small   ○ Normal   ○ Large

1) $4\pi i$
2) $2\pi i$
3) 0
4) $-2\pi i$

7.16. Calculate the integral below. Herein, the direction of contour is counterclockwise.

$$I = \oint_C \frac{z^2}{\sin z}, \quad C : |z| = 8$$

Difficulty level   ○ Easy   ○ Normal   ● Hard

Calculation amount   ○ Small   ○ Normal   ● Large
1) 0
2) $4\pi i$
3) $6\pi^2 i$
4) $12\pi^3 i$

7.17. Calculate the integral below where the contour C is shown in Fig. 7.1.

$$I = \oint_C \frac{z^3 + 3}{z(z-i)^2}$$

Difficulty level   ○ Easy   ○ Normal   ● Hard
Calculation amount   ○ Small   ● Normal   ○ Large
1) $4\pi(-1 + 3i)$
2) $2\pi(1 + 3i)$
3) $2\pi(-1 + 3i)$
4) $3\pi(-1 + 4i.)$

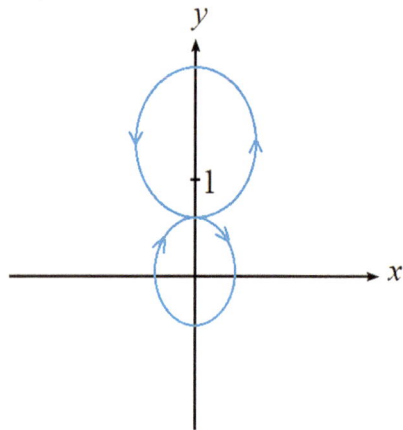

Fig. 7.1. The contour related to Problem 7.17

Exercise

Calculate the integral below where the contour C is shown in Fig. 7.2.

$$I = \oint_C \frac{zdz}{(z^2 - 4)}$$

Final Answer
0

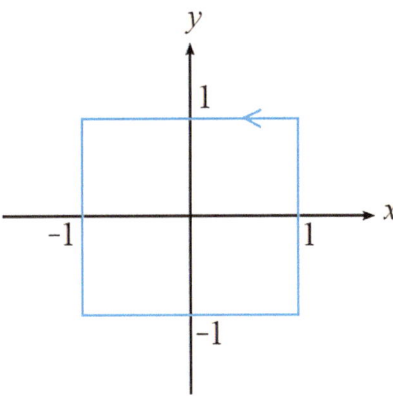

Fig. 7.2. The contour related to the exercise

7.18. Calculate the following integral: Herein, the direction of contour is counterclockwise.

$$I = \frac{3}{\pi i} \oint_{|z|=3} \frac{z^5}{(z-2)^4} \, dz$$

Difficulty level   ○ Easy   ● Normal   ○ Hard
Calculation amount   ○ Small   ● Normal   ○ Large
1) 300
2) 240
3) 120
4) 60

Exercise

Calculate the following integral: Herein, the direction of contour is clockwise.

$$I = \oint_{|z|=2} \frac{dz}{z^4 - 1}$$

Final Answer
()

References

1. Rahmani-Andebili, M. (2024). Precalculus (2nd Ed.) – Practice Problems, Methods, and Solutions, Springer Nature.
2. Rahmani-Andebili, M. (2023). Calculus III – Practice Problems, Methods, and Solutions, Springer Nature.
3. Rahmani-Andebili, M. (2023). Calculus II – Practice Problems, Methods, and Solutions, Springer Nature.
4. Rahmani-Andebili, M. (2023). Calculus I (2nd Ed.) – Practice Problems, Methods, and Solutions, Springer Nature.
5. Rahmani-Andebili, M. (2022). Differential Equations – Practice Problems, Methods, and Solutions, Springer Nature.
6. Rahmani-Andebili, M. (2021). Calculus – Practice Problems, Methods, and Solutions, Springer Nature.
7. Rahmani-Andebili, M. (2021). Precalculus – Practice Problems, Methods, and Solutions, Springer Nature.

Complex Integration: Solutions of Problems

8

Abstract

In this chapter, the problems of the seventh chapter are fully solved, in detail, step-by-step, and with different methods.

8.1 Complex Integration of Nonholomorphic Functions

8.1. Based on the information given in the problem, we have [1–7]:

$$I = \int_C \bar{z} \, dz \tag{1}$$

$$C : \begin{cases} x = t^2 \\ y = t \end{cases} \tag{2}$$

$$0 \le t \le 2 \tag{3}$$

As we know, $z = x + iy$ and $\bar{z} = x - iy$. Thus, from (2), we have:

$$z = t^2 + it \Rightarrow dz = (2t + i)dt \tag{4}$$

$$\bar{z} = t^2 - it \tag{5}$$

Solving (1) and (3), (4), and (5):

$$I = \int_0^2 \left(t^2 - it \right)(2t + i)dt = \int_0^2 \left(2t^3 - it^2 + t \right)dt$$

$$\Rightarrow I = 2\left[\frac{t^4}{4} \right]_0^2 - i\left[\frac{t^3}{3} \right]_0^2 + \left[\frac{t^2}{2} \right]_0^2 = \frac{2^4 - 0}{2} - i\frac{2^3 - 0}{3} + \frac{2^2 - 0}{2}$$

$$\Rightarrow I = 10 - \frac{8}{3}i$$

Choice (4) is the answer.

M. Rahmani-Andebili, *Mathematics of Engineering and Science*, https://doi.org/10.1007/978-3-031-71934-9_8

Notes

In this problem, the relation below has been used:

$$\int x^n dx = \frac{1}{n+1} x^{n+1} + c$$

8.2. Based on the information given in the problem, we have:

$$I = \oint_C z d\bar{z} \tag{1}$$

$$C : z = (1+i) + e^{i\theta}, \quad 0 \le \theta < 2\pi \tag{2}$$

From (2), we have:

$$\bar{z} = 1 - i + e^{-i\theta} \tag{3}$$

$$\Rightarrow d\bar{z} = -ie^{-i\theta} d\theta \tag{4}$$

Solving (1), (2), (3), and (4):

$$I = \int_0^{2\pi} \left(1 + i + e^{i\theta}\right)\left(-ie^{-i\theta}\right) d\theta \tag{5}$$

$$\Rightarrow I = \int_0^{2\pi} \left[(1-i)e^{-i\theta} - i\right] d\theta \tag{6}$$

$$\Rightarrow I = (1-i)\left[-\frac{1}{i} e^{-i\theta}\right]_0^{2\pi} - i[\theta]_0^{2\pi} \tag{7}$$

$$\Rightarrow I = -\frac{1-i}{i}\left(e^{-i2\pi} - e^0\right) - i(2\pi) \tag{8}$$

$$\Rightarrow I = -2\pi i$$

Choice (2) is the answer.

Notes

In this problem, the relations below have been used:

The equation of a circle in Cartesian and polar coordinates with the radius r and the center at z_0 are as follows, respectively:

$$|z - z_0| = r$$

$$z = z_0 + re^{i\theta}, \quad 0 \le \theta < 2\pi$$

$$\frac{d}{dx}e^{ax} = ae^{ax}$$

$$\int e^{ax}dx = \frac{e^{ax}}{a} + c$$

$$\int x^n dx = \frac{1}{n+1}x^{n+1} + c$$

$$e^{-i2\pi} = \cos(-2\pi) + i\sin(-2\pi) = 1$$

$$e^0 = 1$$

8.2 Complex Integration of Holomorphic Functions

8.3. Based on the information given in the problem, we know that the direction of contour is counterclockwise. In addition:

$$I = \oint_C \frac{z+i}{z(z+2)}$$

$$C : |z+1-2i| = 1$$

The singular points of the complex function can be calculated as follows:

$$z(z+2) \Rightarrow z = 0, \, -2$$

Now, we need to see if these points are inside the circle or not.

$$z = 0 \Rightarrow |0+1-2i| = |1-2i| = \sqrt{(1)^2 + (-2)^2} = \sqrt{5} > 1$$

$$z = -2 \Rightarrow |-2+1-2i| = |-1-2i| = \sqrt{(-1)^2 + (-2)^2} = \sqrt{5} > 1$$

Therefore, none of the singular points is inside the contour. Hence, the function under the integral is holomorphic everywhere inside the contour.

Cauchy's integral theorem states that the integral around the contour C is zero if $f(z)$ on and inside the loop is holomorphic everywhere. In other words:

$$I = \oint_C f(z)dz = 0$$

Thus:

$$I = 0$$

Choice (1) is the answer.

Notes

In this problem, the relations below have been used:

$$|a + ib| = \sqrt{a^2 + b^2}$$

The singular points of the complex function $f(z) = \frac{p(z)}{q(z)}$ can be determined by finding the roots of the equation $q(z) = 0$.

For a circle with equation $|z - z_0| = R$, the point z_1 is inside the circle, on the circle, and outside the circle, respectively, if:

$$|z_1 - z_0| < R$$

$$|z_1 - z_0| = R$$

$$|z_1 - z_0| > R$$

8.4. Based on the information given in the problem, we know that the direction of contour is counterclockwise. In addition:

$$I = \oint_{C:|z|=1} f(z) dz$$

$$f(z) = z^3 \bar{z} \cos z$$

Although the complex function seems to be nonholomorphic because of the term $\bar{z}$, it is holomorphic as can be seen below:

$$f(z) = z^3 \bar{z} \cos z = z^2 z \bar{z} \cos z = z^2 |z| \cos z$$

From the contour, we know that $|z| = 1$. Therefore:

$$f(z) = z^2 \cos z$$

As can be seen, the complex function is holomorphic everywhere.

Cauchy's integral theorem states that the integral around the contour C is zero if $f(z)$ on and inside the loop is holomorphic everywhere. In other words:

$$I = \oint_C f(z) dz = 0$$

Thus:

$$I = 0$$

Choice (1) is the answer.

In this problem, the relation below has been used:

$$z\bar{z} = |z|$$

8.3   Complex Integration of Functions Including Finite Number of Singular Points

8.5. Based on the information given in the problem, we know that the contour C is a counterclockwise simple closed loop that surrounds the origin, and we have:

$$I = \oint_C \frac{dz}{z}$$

If $f(z)$ is holomorphic on and inside the counterclockwise contour C (a closed loop that does not intersect itself) except at the finite number of singular points $z_1, z_2, \ldots, z_n$ and the contour surrounds these singular points, then the contour integral can be calculated as follows:

$$I = \oint_C f(z)dz = 2\pi i \left(\operatorname*{Res}_{z=z_1} f(z) + \operatorname*{Res}_{z=z_2} f(z) + \ldots + \operatorname*{Res}_{z=z_n} f(z) \right)$$

Moreover, if the direction of the contour C is clockwise, the contour integral is calculated as follows:

$$I = \oint_C f(z)dz = -2\pi i \left(\operatorname*{Res}_{z=z_1} f(z) + \operatorname*{Res}_{z=z_2} f(z) + \ldots + \operatorname*{Res}_{z=z_n} f(z) \right)$$

The singular point $z = 0$ is the only singular point of the function which is surrounded by the contour. Therefore:

$$I = 2\pi i \left[\operatorname*{Res}_{z=0} f(z) \right]$$

$$\Rightarrow I = 2\pi i \left[\lim_{z \to 0} (z - 0) \frac{1}{z} \right]$$

$$\Rightarrow I = 2\pi i \left[\lim_{z \to 0} 1 \right]$$

$$\Rightarrow I = 2\pi i$$

Choice (2) is the answer.

In this problem, the relations below have been used:

The singular points of the complex function $f(z) = \frac{p(z)}{q(z)}$ can be determined by finding the roots of the equation $q(z) = 0$.

The residue of the function $f(z)$ at the given singularity can be calculated by using two methods as follows:

Method 1: The coefficient of $\frac{1}{z-z_0}$ in the Taylor series expansion of the function $f(z)$ at the essential singular point $z = z_0$ as well as the coefficient of $\frac{1}{z}$ in the Laurent series expansion of the function $f(z)$ at the essential singular point $z = 0$ is called residue.

Method 2: If z_0 is the mth-order pole of a complex function $f(z)$, then the residue of the function at $z = z_0$ can be calculated as follows:

$$\operatorname*{Res}_{z=z_0} f(z) = \frac{1}{(m-1)!} \lim_{z \to z_0} \frac{d^{m-1}}{dz^{m-1}} [(z-z_0)^m f(z)]$$

If z_0 is the first-order pole of a complex function $f(z)$, then the residue of the function at $z = z_0$ can be calculated as follows:

$$\operatorname*{Res}_{z=z_0} f(z) = \lim_{z \to z_0} (z-z_0) f(z)$$

8.6. Based on the information given in the problem, we know that the contour C is a clockwise simple closed loop that surrounds $z = 1$. Moreover:

$$I = \oint_C \frac{\cos \pi z}{z-1}$$

If $f(z)$ is holomorphic on and inside the counterclockwise contour C (a closed loop that does not intersect itself) except at the finite number of singular points $z_1, z_2, \ldots, z_n$ and the contour surrounds these singular points, then the contour integral can be calculated as follows:

$$I = \oint_C f(z) dz = 2\pi i \left(\operatorname*{Res}_{z=z_1} f(z) + \operatorname*{Res}_{z=z_2} f(z) + \ldots + \operatorname*{Res}_{z=z_n} f(z) \right)$$

Moreover, if the direction of the contour C is clockwise, the contour integral is calculated as follows:

$$I = \oint_C f(z) dz = -2\pi i \left(\operatorname*{Res}_{z=z_1} f(z) + \operatorname*{Res}_{z=z_2} f(z) + \ldots + \operatorname*{Res}_{z=z_n} f(z) \right)$$

The singular point of the complex function can be calculated as follows:

$$z - 1 = 0 \Rightarrow z = 1$$

Therefore:

$$I = -2\pi i \left[\operatorname*{Res}_{z=1} f(z) \right]$$

$$\Rightarrow I = -2\pi i \left[\lim_{z \to 1} (z-1) \frac{\cos \pi z}{z-1} \right]$$

$$\Rightarrow I = -2\pi i \left[\lim_{z \to 1} \cos \pi z \right] = -2\pi i \times \cos \pi$$

$$\Rightarrow I = 2\pi i$$

Choice (2) is the answer.

Notes

In this problem, the relations below have been used:

The singular points of the complex function $f(z) = \frac{p(z)}{q(z)}$ can be determined by finding the roots of the equation $q(z) = 0$.

The residue of the function $f(z)$ at the given singularity can be calculated by using two methods as follows:

Method 1: The coefficient of $\frac{1}{z-z_0}$ in the Taylor series expansion of the function $f(z)$ at the essential singular point $z = z_0$ as well as the coefficient of $\frac{1}{z}$ in the Laurent series expansion of the function $f(z)$ at the essential singular point $z = 0$ is called residue.

Method 2: If z_0 is the mth-order pole of a complex function $f(z)$, then the residue of the function at $z = z_0$ can be calculated as follows:

$$\operatorname*{Res}_{z=z_0} f(z) = \frac{1}{(m-1)!} \lim_{z \to z_0} \frac{d^{m-1}}{dz^{m-1}} [(z-z_0)^m f(z)]$$

If z_0 is the first-order pole of a complex function $f(z)$, then the residue of the function at $z = z_0$ can be calculated as follows:

$$\operatorname*{Res}_{z=z_0} f(z) = \lim_{z \to z_0} (z-z_0) f(z)$$

$$\cos \pi = -1$$

8.7. Based on the information given in the problem, we know that the direction of contour is clockwise. In addition:

$$I = \oint_{|z|=2} \frac{\cosh iz}{z^2 + 4z + 3}$$

If $f(z)$ is holomorphic on and inside the counterclockwise contour C (a closed loop that does not intersect itself) except at the finite number of singular points $z_1, z_2, \ldots, z_n$ and the contour surrounds these singular points, then the contour integral can be calculated as follows:

$$I = \oint_C f(z) dz = 2\pi i \left(\operatorname*{Res}_{z=z_1} f(z) + \operatorname*{Res}_{z=z_2} f(z) + \ldots + \operatorname*{Res}_{z=z_n} f(z) \right)$$

Moreover, if the direction of the contour C is clockwise, the contour integral is calculated as follows:

$$I = \oint_C f(z) dz = -2\pi i \left(\operatorname*{Res}_{z=z_1} f(z) + \operatorname*{Res}_{z=z_2} f(z) + \ldots + \operatorname*{Res}_{z=z_n} f(z) \right)$$

The singular points of the complex function can be calculated as follows:

$$z^2 + 4z + 3 = 0 \Rightarrow z = -1, -3$$

Now, we need to see if these singular points are inside the circle or not.

$$z = -1 \Rightarrow |-1| = 1 < 2$$

$$z = -3 \Rightarrow |-3| = 3 > 2$$

As can be seen, only the singular point $z = -1$ is inside the contour. Therefore:

$$I = -2\pi i \left[\operatorname*{Res}_{z=-1} f(z) \right]$$

$$\Rightarrow I = -2\pi i \left[\lim_{z \to -1} (z - (-1)) \frac{\cosh iz}{z^2 + 4z + 3} \right]$$

$$\Rightarrow I = -2\pi i \left[\lim_{z \to -1} \frac{\cosh iz}{z + 3} \right]$$

$$\Rightarrow I = -2\pi i \left[\frac{\cosh(-i)}{-1 + 3} \right]$$

$$\Rightarrow I = -\pi i \cos 1$$

Choice (2) is the answer.

Notes

In this problem, the relations below have been used:

The singular points of the complex function $f(z) = \frac{p(z)}{q(z)}$ can be determined by finding the roots of the equation $q(z) = 0$.

For a circle with equation $|z - z_0| = R$, the point z_1 is inside the circle, on the circle, and outside the circle, respectively, if:

$$|z_1 - z_0| < R$$

$$|z_1 - z_0| = R$$

$$|z_1 - z_0| > R$$

The residue of the function $f(z)$ at the given singularity can be calculated by using two methods as follows:

Method 1: The coefficient of $\frac{1}{z - z_0}$ in the Taylor series expansion of the function $f(z)$ at the essential singular point $z = z_0$ as well as the coefficient of $\frac{1}{z}$ in the Laurent series expansion of the function $f(z)$ at the essential singular point $z = 0$ is called residue.

Method 2: If z_0 is the mth-order pole of a complex function $f(z)$, then the residue of the function at $z = z_0$ can be calculated as follows:

$$\operatorname*{Res}_{z=z_0} f(z) = \frac{1}{(m-1)!} \lim_{z \to z_0} \frac{d^{m-1}}{dz^{m-1}} [(z - z_0)^m f(z)]$$

If z_0 is the first-order pole of a complex function $f(z)$, then the residue of the function at $z = z_0$ can be calculated as follows:

$$\operatorname*{Res}_{z=z_0} f(z) = \lim_{z \to z_0} (z - z_0) f(z)$$

$$\cosh(-z) = \cosh z$$

$$\cosh iz = \cos z$$

8.8. Based on the information given in the problem, we know that the direction of contour is counterclockwise. In addition:

$$I = \oint_C \frac{z+1}{z^3 - 2z^2} dz$$

$$C : |z - 2 - i| = 2$$

If $f(z)$ is holomorphic on and inside the counterclockwise contour C (a closed loop that does not intersect itself) except at the finite number of singular points $z_1, z_2, \ldots, z_n$ and the contour surrounds these singular points, then the contour integral can be calculated as follows:

$$I = \oint_C f(z) dz = 2\pi i \left(\operatorname*{Res}_{z=z_1} f(z) + \operatorname*{Res}_{z=z_2} f(z) + \ldots + \operatorname*{Res}_{z=z_n} f(z) \right)$$

Moreover, if the direction of the contour C is clockwise, the contour integral is calculated as follows:

$$I = \oint_C f(z) dz = -2\pi i \left(\operatorname*{Res}_{z=z_1} f(z) + \operatorname*{Res}_{z=z_2} f(z) + \ldots + \operatorname*{Res}_{z=z_n} f(z) \right)$$

The singular points of the complex function can be calculated as follows:

$$z^3 - 2z^2 = 0 \Rightarrow z = 0, 2$$

However, only the singular points $z = 2$ is inside the contour $|z - 2 - i| = 2$ as can be seen in the following:

$$z = 0 \Rightarrow |0 - 2 - i| = |-2 - i| = \sqrt{5} > 2$$

$$z = 2 \Rightarrow |2 - 2 - i| = |-i| = 1 < 2$$

As can be seen, only the singular point $z = 2$ is inside the contour. Therefore:

$$I = 2\pi i \left[\operatorname*{Res}_{z=2} f(z) \right]$$

$$\Rightarrow I = 2\pi i \left[\lim_{z \to 2} (z - 2) \frac{z+1}{z^3 - 2z^2} \right]$$

$$\Rightarrow I = 2\pi i \left[\lim_{z \to 2} \frac{z+1}{z^2} \right] = 2\pi i \times \frac{3}{4}$$

$$\Rightarrow I = \frac{3}{2}\pi i$$

Choice (4) is the answer.

Notes

In this problem, the relations below have been used:

The singular points of the complex function $f(z) = \frac{p(z)}{q(z)}$ can be determined by finding the roots of the equation $q(z) = 0$.

For a circle with equation $|z - z_0| = R$, the point z_1 is inside the circle, on the circle, and outside the circle, respectively, if:

$$|z_1 - z_0| < R$$

$$|z_1 - z_0| = R$$

$$|z_1 - z_0| > R$$

The residue of the function $f(z)$ at the given singularity can be calculated by using two methods as follows:

Method 1: The coefficient of $\frac{1}{z - z_0}$ in the Taylor series expansion of the function $f(z)$ at the essential singular point $z = z_0$ as well as the coefficient of $\frac{1}{z}$ in the Laurent series expansion of the function $f(z)$ at the essential singular point $z = 0$ is called residue.

Method 2: If z_0 is the mth-order pole of a complex function $f(z)$, then the residue of the function at $z = z_0$ can be calculated as follows:

$$\operatorname*{Res}_{z=z_0} f(z) = \frac{1}{(m-1)!} \lim_{z \to z_0} \frac{d^{m-1}}{dz^{m-1}} \left[(z - z_0)^m f(z) \right]$$

If z_0 is the first-order pole of a complex function $f(z)$, then the residue of the function at $z = z_0$ can be calculated as follows:

$$\operatorname*{Res}_{z=z_0} f(z) = \lim_{z \to z_0} (z - z_0) f(z)$$

8.9. Based on the information given in the problem, we know that the contour C is a counterclockwise square with the corners located at ± 4 and $\pm 4i$. Also, we have:

$$I = \oint_C \frac{e^z}{z^4} dz$$

If $f(z)$ is holomorphic on and inside the counterclockwise contour C (a closed loop that does not intersect itself) except at the finite number of singular points $z_1, z_2, \ldots, z_n$ and the contour surrounds these singular points, then the contour integral can be calculated as follows:

$$I = \oint_C f(z)dz = 2\pi i \left(\operatorname*{Res}_{z=z_1} f(z) + \operatorname*{Res}_{z=z_2} f(z) + \ldots + \operatorname*{Res}_{z=z_n} f(z) \right)$$

Moreover, if the direction of the contour C is clockwise, the contour integral is calculated as follows:

$$I = \oint_C f(z)dz = -2\pi i \left(\operatorname*{Res}_{z=z_1} f(z) + \operatorname*{Res}_{z=z_2} f(z) + \ldots + \operatorname*{Res}_{z=z_n} f(z) \right)$$

The singular points of the complex function can be calculated as follows:

$$z^4 = 0 \Rightarrow z = 0$$

It is seen that the singular point, which is a four-order pole, is inside the contour. Therefore:

$$I = 2\pi i \left[\operatorname*{Res}_{z=0} f(z) \right]$$

$$\Rightarrow I = 2\pi i \left[\frac{1}{3!} \lim_{z \to 0} \frac{d^3}{dz^3} \left[(z-0)^4 \frac{e^z}{z^4} \right] \right]$$

$$\Rightarrow I = \frac{\pi i}{3} \left[\lim_{z \to 0} \frac{d^3}{dz^3} [e^z] \right]$$

$$\Rightarrow I = \frac{\pi i}{3} \left[\lim_{z \to 0} e^z \right]$$

$$\Rightarrow I = \frac{\pi i}{3}$$

Choice (2) is the answer.

Notes

In this problem, the relations below have been used:

$$\frac{d}{dz} e^z = e^z$$

The singular points of the complex function $f(z) = \frac{p(z)}{q(z)}$ can be determined by finding the roots of the equation $q(z) = 0$.

The residue of the function $f(z)$ at the given singularity can be calculated by using two methods as follows:

Method 1: The coefficient of $\frac{1}{z-z_0}$ in the Taylor series expansion of the function $f(z)$ at the essential singular point $z = z_0$ as well as the coefficient of $\frac{1}{z}$ in the Laurent series expansion of the function $f(z)$ at the essential singular point $z = 0$ is called residue.

Method 2: If z_0 is the mth-order pole of a complex function $f(z)$, then the residue of the function at $z = z_0$ can be calculated as follows:

$$\operatorname*{Res}_{z=z_0} f(z) = \frac{1}{(m-1)!} \lim_{z \to z_0} \frac{d^{m-1}}{dz^{m-1}} \left[(z - z_0)^m f(z) \right]$$

If z_0 is the first-order pole of a complex function $f(z)$, then the residue of the function at $z = z_0$ can be calculated as follows:

$$\operatorname*{Res}_{z=z_0} f(z) = \lim_{z \to z_0} (z - z_0) f(z)$$

8.10. Based on the information given in the problem, we know that the direction of contour is clockwise. In addition:

$$I = \oint_{|z-1|=1} \frac{\sin \pi z}{(z^2 - 1)^2}$$

If $f(z)$ is holomorphic on and inside the counterclockwise contour C (a closed loop that does not intersect itself) except at the finite number of singular points $z_1, z_2, \ldots, z_n$ and the contour surrounds these singular points, then the contour integral can be calculated as follows:

$$I = \oint_C f(z) dz = 2\pi i \left(\operatorname*{Res}_{z=z_1} f(z) + \operatorname*{Res}_{z=z_2} f(z) + \ldots + \operatorname*{Res}_{z=z_n} f(z) \right)$$

Moreover, if the direction of the contour C is clockwise, the contour integral is calculated as follows:

$$I = \oint_C f(z) dz = -2\pi i \left(\operatorname*{Res}_{z=z_1} f(z) + \operatorname*{Res}_{z=z_2} f(z) + \ldots + \operatorname*{Res}_{z=z_n} f(z) \right)$$

The singular points of the complex function can be calculated as follows:

$$\left(z^2 - 1 \right)^2 = 0 \Rightarrow z = -1, 1$$

Now, we need to see if the singular point is inside the circle or not.

$$z = -1 \Rightarrow |(-1) - 1| = 2 > 1$$

$$z = 1 \Rightarrow |1 - 1| = 0 < 1$$

As can be seen, only the singular point $z = 1$, which is a second-order pole, is inside the contour. Thus:

$$I = -2\pi i \left[\operatorname*{Res}_{z=1} f(z) \right]$$

$$\Rightarrow I = -2\pi i \left[\lim_{z \to 1} \frac{d}{dz} \left[(z - 1)^2 \frac{\sin \pi z}{(z^2 - 1)^2} \right] \right]$$

$$\Rightarrow I = -2\pi i \left[\lim_{z \to 1} \frac{d}{dz} \left[\frac{\sin \pi z}{(z + 1)^2} \right] \right]$$

$$\Rightarrow I = -2\pi i \left[\lim_{z \to 1} \left[\frac{\pi \cos \pi z (z+1)^2 - 2(z+1) \sin \pi z}{(z+1)^4} \right] \right]$$

$$\Rightarrow I = -2\pi i \left[\frac{4\pi \cos \pi - 4 \sin \pi}{16} \right]$$

$$\Rightarrow I = \frac{\pi^2}{2} i$$

Choice (4) is the answer.

Notes

In this problem, the relations below have been used:

The singular points of the complex function $f(z) = \frac{p(z)}{q(z)}$ can be determined by finding the roots of the equation $q(z) = 0$.

For a circle with equation $|z - z_0| = R$, the point z_1 is inside the circle, on the circle, and outside the circle, respectively, if:

$$|z_1 - z_0| < R$$

$$|z_1 - z_0| = R$$

$$|z_1 - z_0| > R$$

The residue of the function $f(z)$ at the given singularity can be calculated by using two methods as follows:

Method 1: The coefficient of $\frac{1}{z - z_0}$ in the Taylor series expansion of the function $f(z)$ at the essential singular point $z = z_0$ as well as the coefficient of $\frac{1}{z}$ in the Laurent series expansion of the function $f(z)$ at the essential singular point $z = 0$ is called residue.

Method 2: If z_0 is the mth-order pole of a complex function $f(z)$, then the residue of the function at $z = z_0$ can be calculated as follows:

$$\operatorname*{Res}_{z=z_0} f(z) = \frac{1}{(m-1)!} \lim_{z \to z_0} \frac{d^{m-1}}{dz^{m-1}} [(z - z_0)^m f(z)]$$

If z_0 is the first-order pole of a complex function $f(z)$, then the residue of the function at $z = z_0$ can be calculated as follows:

$$\operatorname*{Res}_{z=z_0} f(z) = \lim_{z \to z_0} (z - z_0) f(z)$$

In this problem, the relations below have been used:

$$\frac{d}{dz} \left(\frac{f(z)}{g(z)} \right) = \frac{f'(z)g(z) - g'(z)f(z)}{(g(z))^2}$$

$$\frac{d}{dz}(\sin az) = a \cos az$$

$$\frac{d}{dz}(f(z))^n = nf'(z)(f(z))^{n-1}$$

$$\cos \pi = -1$$

$$\sin \pi = 0$$

$$i^2 = 1$$

8.11. Based on the information given in the problem, we know that the direction of contour is counterclockwise. In addition:

$$I = \oint_{|z|=1} \left(z + \frac{1}{z} \right) e^{\frac{1}{z}} dz$$

If $f(z)$ is holomorphic on and inside the counterclockwise contour C (a closed loop that does not intersect itself) except at the finite number of singular points $z_1, z_2, \ldots, z_n$ and the contour surrounds these singular points, then the contour integral can be calculated as follows:

$$I = \oint_C f(z)dz = 2\pi i \left(\operatorname*{Res}_{z=z_1} f(z) + \operatorname*{Res}_{z=z_2} f(z) + \ldots + \operatorname*{Res}_{z=z_n} f(z) \right)$$

Moreover, if the direction of the contour C is clockwise, the contour integral is calculated as follows:

$$I = \oint_C f(z)dz = -2\pi i \left(\operatorname*{Res}_{z=z_1} f(z) + \operatorname*{Res}_{z=z_2} f(z) + \ldots + \operatorname*{Res}_{z=z_n} f(z) \right)$$

By replacing $e^{\frac{1}{z}}$ by its Laurent series expansion, we have:

$$\left(z + \frac{1}{z} \right) e^{\frac{1}{z}} = \left(z + \frac{1}{z} \right) \left(1 + \frac{1}{z} + \frac{1}{2!z^2} + \frac{1}{3!z^3} + \frac{1}{4!z^4} + \cdots \right)$$

As can be seen, $z = 0$ is the essential singular point of the function because its Laurent series expansion includes an infinite number of terms with a negative exponent of z. Moreover, the singular point resides inside the contour $|z| = 1$, as can be seen in the following:

$$z = 0 \Rightarrow |0| = 0 < 1$$

The coefficient of $\frac{1}{z}$ as the residue of the function at $z = 0$ can be calculated as follows:

$$\operatorname*{Res}_{z=0} f(z) = 1 + \frac{1}{2!} = \frac{3}{2}$$

Therefore:

$$I = 2\pi i \times \frac{3}{2}$$

$$\Rightarrow I = 3\pi i$$

Choice (4) is the answer.

Notes

In this problem, the relations below have been used:

The Laurent series expansion of e^z is as follows:

$$e^z = \sum_{n=0}^{\infty} \frac{z^n}{n!} = 1 + z + \frac{z^2}{2!} + \cdots, \quad |z| < \infty$$

For a circle with equation $|z - z_0| = R$, the point z_1 is inside the circle, on the circle, and outside the circle, respectively, if:

$$|z_1 - z_0| < R$$

$$|z_1 - z_0| = R$$

$$|z_1 - z_0| > R$$

The residue of the function $f(z)$ at the given singularity can be calculated by using two methods as follows:

Method 1: The coefficient of $\frac{1}{z - z_0}$ in the Taylor series expansion of the function $f(z)$ at the essential singular point $z = z_0$ as well as the coefficient of $\frac{1}{z}$ in the Laurent series expansion of the function $f(z)$ at the essential singular point $z = 0$ is called residue.

Method 2: If z_0 is the mth-order pole of a complex function $f(z)$, then the residue of the function at $z = z_0$ can be calculated as follows:

$$\operatorname*{Res}_{z=z_0} f(z) = \frac{1}{(m-1)!} \lim_{z \to z_0} \frac{d^{m-1}}{dz^{m-1}} \left[(z - z_0)^m f(z) \right]$$

If z_0 is the first-order pole of a complex function $f(z)$, then the residue of the function at $z = z_0$ can be calculated as follows:

$$\operatorname*{Res}_{z=z_0} f(z) = \lim_{z \to z_0} (z - z_0) f(z)$$

8.12. Based on the information given in the problem, we know that the contour C is a counterclockwise square with the corners located at $\pm\sqrt{2}$ and $\pm i\sqrt{2}$. Moreover:

$$I = \oint_C \frac{\cosh \pi z}{z(z^2 + 1)}$$

If $f(z)$ is holomorphic on and inside the counterclockwise contour C (a closed loop that does not intersect itself) except at the finite number of singular points $z_1, z_2, \ldots, z_n$ and the contour surrounds these singular points, then the contour integral can be calculated as follows:

$$I = \oint_C f(z)dz = 2\pi i \left(\operatorname*{Res}_{z=z_1} f(z) + \operatorname*{Res}_{z=z_2} f(z) + \ldots + \operatorname*{Res}_{z=z_n} f(z) \right)$$

Moreover, if the direction of the contour C is clockwise, the contour integral is calculated as follows:

$$I = \oint_C f(z)dz = -2\pi i \left(\operatorname*{Res}_{z=z_1} f(z) + \operatorname*{Res}_{z=z_2} f(z) + \ldots + \operatorname*{Res}_{z=z_n} f(z) \right)$$

The singular points of the complex function can be calculated as follows:

$$z(z^2 + 1) = 0 \Rightarrow z = 0, -i, i$$

All these singular points are inside the contour. Therefore:

$$I = 2\pi i \left[\operatorname*{Res}_{z=0} f(z) + \operatorname*{Res}_{z=-i} f(z) + \operatorname*{Res}_{z=i} f(z) \right]$$

$$\Rightarrow I = 2\pi i \left[\lim_{z \to 0} (z - 0) \frac{\cosh \pi z}{z(z^2 + 1)} + \lim_{z \to -i} (z - (-i)) \frac{\cosh \pi z}{z(z^2 + 1)} + \lim_{z \to i} (z - i) \frac{\cosh \pi z}{z(z^2 + 1)} \right]$$

$$\Rightarrow I = 2\pi i \left[\lim_{z \to 0} \frac{\cosh \pi z}{z^2 + 1} + \lim_{z \to -i} \frac{\cosh \pi z}{z(z - i)} + \lim_{z \to i} \frac{\cosh \pi z}{z(z + i)} \right]$$

$$\Rightarrow I = 2\pi i \left[\frac{\cosh 0}{0 + 1} + \frac{\cosh(-i\pi)}{-i(-i - i)} + \frac{\cosh i\pi}{(i)(i + i)} \right]$$

$$\Rightarrow I = 2\pi i \left[\cosh 0 + \frac{\cos \pi}{2i^2} + \frac{\cos \pi}{2i^2} \right]$$

$$\Rightarrow I = 2\pi i \left[1 + \frac{-1}{2i^2} + \frac{-1}{2i^2} \right]$$

$$\Rightarrow I = 4\pi i$$

Choice (3) is the answer.

Notes

In this problem, the relations below have been used:

The singular points of the complex function $f(z) = \frac{p(z)}{q(z)}$ can be determined by finding the roots of the equation $q(z) = 0$.

The residue of the function $f(z)$ at the given singularity can be calculated by using two methods as follows:

Method 1: The coefficient of $\frac{1}{z - z_0}$ in the Taylor series expansion of the function $f(z)$ at the essential singular point $z = z_0$ as well as the coefficient of $\frac{1}{z}$ in the Laurent series expansion of the function $f(z)$ at the essential singular point $z = 0$ is called residue.

Method 2: If z_0 is the mth-order pole of a complex function $f(z)$, then the residue of the function at $z = z_0$ can be calculated as follows:

$$\operatorname{Res}_{z=z_0} f(z) = \frac{1}{(m-1)!} \lim_{z \to z_0} \frac{d^{m-1}}{dz^{m-1}} [(z-z_0)^m f(z)]$$

If z_0 is the first-order pole of a complex function $f(z)$, then the residue of the function at $z = z_0$ can be calculated as follows:

$$\operatorname{Res}_{z=z_0} f(z) = \lim_{z \to z_0} (z-z_0) f(z)$$

$$\cosh 0 = 1$$

$$\cosh(iz) = \cos z$$

$$\cos(-z) = \cos z$$

$$\cos \pi = -1$$

$$i^2 = -1$$

8.13. Based on the information given in the problem, we know that the direction of contour is counterclockwise. In addition:

$$I = \oint_{|z|=2} \tan z \, dz$$

$$\Rightarrow I = \oint_{|z|=2} \frac{\sin z}{\cos z} dz$$

If $f(z)$ is holomorphic on and inside the counterclockwise contour C (a closed loop that does not intersect itself) except at the finite number of singular points z_1, z_2, ..., z_n and the contour surrounds these singular points, then the contour integral can be calculated as follows:

$$I = \oint_C f(z) dz = 2\pi i \left(\operatorname{Res}_{z=z_1} f(z) + \operatorname{Res}_{z=z_2} f(z) + \ldots + \operatorname{Res}_{z=z_n} f(z) \right)$$

Moreover, if the direction of the contour C is clockwise, the contour integral is calculated as follows:

$$I = \oint_C f(z) dz = -2\pi i \left(\operatorname{Res}_{z=z_1} f(z) + \operatorname{Res}_{z=z_2} f(z) + \ldots + \operatorname{Res}_{z=z_n} f(z) \right)$$

The singular points of the complex function can be calculated as follows:

$$\cos z = 0 \Rightarrow z = k\pi \pm \frac{\pi}{2}, k = 0, \pm 1, \pm 2, \cdots$$

Only the singular points $z = -\frac{\pi}{2}, \frac{\pi}{2}$ are inside the contour $|z| = 2$ as can be seen in the following:

$$z = -\frac{\pi}{2} \Rightarrow \left| -\frac{\pi}{2} \right| = \frac{\pi}{2} < 2$$

$$z = \frac{\pi}{2} \Rightarrow \left| \frac{\pi}{2} \right| = \frac{\pi}{2} < 2$$

Therefore:

$$I = 2\pi i \left[\operatorname*{Res}_{z=-\frac{\pi}{2}} f(z) + \operatorname*{Res}_{z=\frac{\pi}{2}} f(z) \right]$$

$$\Rightarrow I = 2\pi i \left[\lim_{z \to -\frac{\pi}{2}} \left(z - \left(-\frac{\pi}{2} \right) \right) \frac{\sin z}{\cos z} + \lim_{z \to \frac{\pi}{2}} \left(z - \frac{\pi}{2} \right) \frac{\sin z}{\cos z} \right]$$

$$\Rightarrow I = 2\pi i \left[\frac{0}{0} + \frac{0}{0} \right]$$

$$^H \Rightarrow I = 2\pi i \left[\lim_{z \to -\frac{\pi}{2}} \frac{\left(\left(z + \frac{\pi}{2} \right) \sin z \right)'}{(\cos z)'} + \lim_{z \to \frac{\pi}{2}} \frac{\left(\left(z - \frac{\pi}{2} \right) \sin z \right)'}{(\cos z)'} \right]$$

$$\Rightarrow I = 2\pi i \left[\lim_{z \to -\frac{\pi}{2}} \frac{\sin z + \left(z + \frac{\pi}{2} \right) \cos z}{- \sin z} + \lim_{z \to \frac{\pi}{2}} \frac{\sin z + \left(z - \frac{\pi}{2} \right) \cos z}{- \sin z} \right]$$

$$\Rightarrow I = 2\pi i \left[\frac{\sin \left(-\frac{\pi}{2} \right) + \left(-\frac{\pi}{2} + \frac{\pi}{2} \right) \cos \left(-\frac{\pi}{2} \right)}{- \sin \left(-\frac{\pi}{2} \right)} + \frac{\sin \left(\frac{\pi}{2} \right) + \left(\frac{\pi}{2} - \frac{\pi}{2} \right) \cos \left(\frac{\pi}{2} \right)}{- \sin \left(\frac{\pi}{2} \right)} \right]$$

$$\Rightarrow I = 2\pi i \left[\frac{-1}{1} + \frac{1}{-1} \right]$$

$$\Rightarrow I = -4\pi i$$

Choice (3) is the answer.

Notes

In this problem, the relations below have been used:

The singular points of the complex function $f(z) = \frac{p(z)}{q(z)}$ can be determined by finding the roots of the equation $q(z) = 0$.

For a circle with equation $|z - z_0| = R$, the point z_1 is inside the circle, on the circle, and outside the circle, respectively, if:

$$|z_1 - z_0| < R$$

$$|z_1 - z_0| = R$$

$$|z_1 - z_0| > R$$

The residue of the function $f(z)$ at the given singularity can be calculated by using two methods as follows:

Method 1: The coefficient of $\frac{1}{z - z_0}$ in the Taylor series expansion of the function $f(z)$ at the essential singular point $z = z_0$ as well as the coefficient of $\frac{1}{z}$ in the Laurent series expansion of the function $f(z)$ at the essential singular point $z = 0$ is called residue.

Method 2: If z_0 is the mth-order pole of a complex function $f(z)$, then the residue of the function at $z = z_0$ can be calculated as follows:

$$\operatorname*{Res}_{z=z_0} f(z) = \frac{1}{(m-1)!} \lim_{z \to z_0} \frac{d^{m-1}}{dz^{m-1}} [(z-z_0)^m f(z)]$$

If z_0 is the first-order pole of a complex function $f(z)$, then the residue of the function at $z = z_0$ can be calculated as follows:

$$\operatorname*{Res}_{z=z_0} f(z) = \lim_{z \to z_0} (z-z_0)f(z)$$

$$h(z) = \lim_{z \to z_0} \frac{f(z)}{g(z)} = \frac{0}{0} \overset{H}{\Rightarrow} h(z) = \lim_{z \to z_0} \frac{f'(z)}{g'(z)}$$

$$\frac{d}{dz}(\sin z) = \cos z$$

$$\frac{d}{dz}(\cos z) = -\sin z$$

$$\sin\left(-\frac{\pi}{2}\right) = -1$$

$$\sin\left(\frac{\pi}{2}\right) = 1$$

8.14. Based on the information given in the problem, we know that the direction of contour is counterclockwise. In addition:

$$I = \oint_C 120 z^3 e^{-\frac{1}{z}} dz$$

$$C : |z| = \frac{1}{3}$$

If $f(z)$ is holomorphic on and inside the counterclockwise contour C (a closed loop that does not intersect itself) except at the finite number of singular points $z_1, z_2, \ldots, z_n$ and the contour surrounds these singular points, then the contour integral can be calculated as follows:

$$I = \oint_C f(z)dz = 2\pi i \left(\operatorname*{Res}_{z=z_1} f(z) + \operatorname*{Res}_{z=z_2} f(z) + \ldots + \operatorname*{Res}_{z=z_n} f(z) \right)$$

Moreover, if the direction of the contour C is clockwise, the contour integral is calculated as follows:

$$I = \oint_C f(z)dz = -2\pi i \left(\operatorname*{Res}_{z=z_1} f(z) + \operatorname*{Res}_{z=z_2} f(z) + \ldots + \operatorname*{Res}_{z=z_n} f(z) \right)$$

By replacing $e^{-\frac{1}{z}}$ by its Laurent series expansion, we have:

$$120 z^3 e^{-\frac{1}{z}} = 120 z^3 \left(1 - \frac{1}{z} + \frac{1}{2! z^2} - \frac{1}{3! z^3} + \frac{1}{4! z^4} - \cdots \right)$$

As can be seen, $z = 0$ is the essential singular point of the function because its Laurent series expansion includes an infinite number of terms with a negative exponent of z. Moreover, the singular point resides inside the contour, as can be seen in the following:

$$z = 0 \Rightarrow |0| = 0 < \frac{1}{3}$$

The coefficient of $\frac{1}{z}$ as the residue of the function at $z = 0$ can be calculated as follows:

$$\operatorname*{Res}_{z=0} f(z) = 120\left(\frac{1}{4!}\right) = \frac{120}{24} = 5$$

Therefore:

$$I = 2\pi i \times 5$$

$$\Rightarrow I = 10\pi i$$

Choice (4) is the answer.

Notes

In this problem, the relations below have been used:

The Laurent series expansion of e^z is as follows:

$$e^z = \sum_{n=0}^{\infty} \frac{z^n}{n!} = 1 + z + \frac{z^2}{2!} + \cdots, |z| < \infty$$

For a circle with equation $|z - z_0| = R$, the point z_1 is inside the circle, on the circle, and outside the circle, respectively, if:

$$|z_1 - z_0| < R$$

$$|z_1 - z_0| = R$$

$$|z_1 - z_0| > R$$

The residue of the function $f(z)$ at the given singularity can be calculated by using two methods as follows:

Method 1: The coefficient of $\frac{1}{z-z_0}$ in the Taylor series expansion of the function $f(z)$ at the essential singular point $z = z_0$ as well as the coefficient of $\frac{1}{z}$ in the Laurent series expansion of the function $f(z)$ at the essential singular point $z = 0$ is called residue.

Method 2: If z_0 is the mth-order pole of a complex function $f(z)$, then the residue of the function at $z = z_0$ can be calculated as follows:

$$\operatorname*{Res}_{z=z_0} f(z) = \frac{1}{(m-1)!} \lim_{z \to z_0} \frac{d^{m-1}}{dz^{m-1}} [(z - z_0)^m f(z)]$$

If z_0 is the first-order pole of a complex function $f(z)$, then the residue of the function at $z = z_0$ can be calculated as follows:

$$\operatorname*{Res}_{z=z_0} f(z) = \lim_{z \to z_0} (z - z_0) f(z)$$

8.15. Based on the information given in the problem, we know that the contour is the unit circle with counterclockwise direction. Moreover:

$$I = \oint_C e^{-\frac{1}{z}} \sin\left(\frac{2}{z}\right) dz$$

If $f(z)$ is holomorphic on and inside the counterclockwise contour C (a closed loop that does not intersect itself) except at the finite number of singular points z_1, z_2, ..., z_n and the contour surrounds these singular points, then the contour integral can be calculated as follows:

$$I = \oint_C f(z)dz = 2\pi i\left(\operatorname*{Res}_{z=z_1} f(z) + \operatorname*{Res}_{z=z_2} f(z) + \ldots + \operatorname*{Res}_{z=z_n} f(z)\right)$$

Moreover, if the direction of the contour C is clockwise, the contour integral is calculated as follows:

$$I = \oint_C f(z)dz = -2\pi i\left(\operatorname*{Res}_{z=z_1} f(z) + \operatorname*{Res}_{z=z_2} f(z) + \ldots + \operatorname*{Res}_{z=z_n} f(z)\right)$$

By replacing $e^{-\frac{1}{z}}$ and $\sin\frac{2}{z}$ by their Laurent series expansions, we have:

$$e^{-\frac{1}{z}} \sin\frac{2}{z} = \left(1 - \frac{1}{z} + \frac{1}{2!z^2} - \frac{1}{3!z^3} + \cdots\right)\left(\frac{2}{z} - \frac{8}{3!z^3} + \cdots\right)$$

As can be seen, $z = 0$ is the essential singular point of the function because its Laurent series expansion includes an infinite number of terms with a negative exponent of z. Moreover, the singular point resides inside the contour, as can be seen in the following:

$$z = 0 \Rightarrow |0| = 0 < \frac{1}{3}$$

The coefficient of $\frac{1}{z}$ as the residue of the function at $z = 0$ can be calculated as follows:

$$\operatorname*{Res}_{z=0} f(z) = 1 \times 2 = 2$$

Therefore:

$$I = 2\pi i \times 2$$

$$\Rightarrow I = 4\pi i$$

Choice (1) is the answer.

Notes

In this problem, the relations below have been used:

The Laurent series expansion of e^z and $\sin z$ are as follows:

$$e^z = \sum_{n=0}^{\infty} \frac{z^n}{n!} = 1 + z + \frac{z^2}{2!} + \cdots, \quad |z| < \infty$$

$$\sin z = \sum (-1)^n \frac{z^{2n+1}}{(2n+1)!} = z - \frac{z^3}{3!} + \frac{z^5}{5!} - \frac{z^7}{7!} + \cdots, \quad |z| < \infty$$

For a circle with equation $|z - z_0| = R$, the point z_1 is inside the circle, on the circle, and outside the circle, respectively, if:

$$|z_1 - z_0| < R$$

$$|z_1 - z_0| = R$$

$$|z_1 - z_0| > R$$

The residue of the function $f(z)$ at the given singularity can be calculated by using two methods as follows:

Method 1: The coefficient of $\frac{1}{z-z_0}$ in the Taylor series expansion of the function $f(z)$ at the essential singular point $z = z_0$ as well as the coefficient of $\frac{1}{z}$ in the Laurent series expansion of the function $f(z)$ at the essential singular point $z = 0$ is called residue.

Method 2: If z_0 is the mth-order pole of a complex function $f(z)$, then the residue of the function at $z = z_0$ can be calculated as follows:

$$\operatorname*{Res}_{z=z_0} f(z) = \frac{1}{(m-1)!} \lim_{z \to z_0} \frac{d^{m-1}}{dz^{m-1}} [(z - z_0)^m f(z)]$$

If z_0 is the first-order pole of a complex function $f(z)$, then the residue of the function at $z = z_0$ can be calculated as follows:

$$\operatorname*{Res}_{z=z_0} f(z) = \lim_{z \to z_0} (z - z_0) f(z)$$

8.16. Based on the information given in the problem, we know that the direction of contour is counterclockwise. In addition:

$$I = \oint_C \frac{z^2}{\sin z}$$

$$C : |z| = 8$$

If $f(z)$ is holomorphic on and inside the counterclockwise contour C (a closed loop that does not intersect itself) except at the finite number of singular points $z_1, z_2, \ldots, z_n$ and the contour surrounds these singular points, then the contour integral can be calculated as follows:

$$I = \oint_C f(z)dz = 2\pi i \left(\operatorname*{Res}_{z=z_1} f(z) + \operatorname*{Res}_{z=z_2} f(z) + \ldots + \operatorname*{Res}_{z=z_n} f(z) \right)$$

Moreover, if the direction of the contour C is clockwise, the contour integral is calculated as follows:

$$I = \oint_C f(z)dz = -2\pi i \left(\operatorname*{Res}_{z=z_1} f(z) + \operatorname*{Res}_{z=z_2} f(z) + \ldots + \operatorname*{Res}_{z=z_n} f(z) \right)$$

The singular points of the complex function can be calculated as follows:

$$\sin z = 0 \Rightarrow z = k\pi, k = 0, \pm 1, \pm 2, \cdots$$

However, only the singular points $z = 0, \pm \pi, \pm 2\pi$ are inside the contour $|z| = 8$ as can be seen in the following:

$$z = 0 \Rightarrow |0| = 0 < 8$$

$$z = \pm \pi \Rightarrow |\pm \pi| = \pi < 8$$

$$z = \pm 2\pi \Rightarrow |\pm 2\pi| = 2\pi < 8$$

Therefore:

$$I = 2\pi i \left[\operatorname*{Res}_{z=0} f(z) + \operatorname*{Res}_{z=-\pi} f(z) + \operatorname*{Res}_{z=\pi} f(z) + \operatorname*{Res}_{z=-2\pi} f(z) + \operatorname*{Res}_{z=2\pi} f(z) \right]$$

$$\Rightarrow I = 2\pi i \left[\lim_{z \to 0}(z-0)\frac{z^2}{\sin z} + \lim_{z \to -\pi}(z-(-\pi))\frac{z^2}{\sin z} + \lim_{z \to \pi}(z-\pi)\frac{z^2}{\sin z} + \lim_{z \to -2\pi}(z-(-2\pi))\frac{z^2}{\sin z} + \lim_{z \to 2\pi}(z-2\pi)\frac{z^2}{\sin z} \right]$$

$$\Rightarrow I = 2\pi i \left[0 + \frac{0}{0} + \frac{0}{0} + \frac{0}{0} + \frac{0}{0} \right]$$

$$H \Rightarrow I = 2\pi i \left[\lim_{z \to 0}\frac{(z^3)'}{(\sin z)'} + \lim_{z \to -\pi}\frac{((z+\pi)z^2)'}{(\sin z)'} + \lim_{z \to \pi}\frac{((z-\pi)z^2)'}{(\sin z)'} + \lim_{z \to -2\pi}\frac{((z+2\pi)z^2)'}{(\sin z)'} + \lim_{z \to 2\pi}\frac{((z-2\pi)z^2)'}{(\sin z)'} \right]$$

$$\Rightarrow I = 2\pi i \left[\lim_{z \to 0}\frac{3z^2}{\cos z} + \lim_{z \to -\pi}\frac{z^2+2(z+\pi)z}{\cos z} + \lim_{z \to \pi}\frac{z^2+2(z-\pi)z}{\cos z} + \lim_{z \to -2\pi}\frac{z^2+2(z+2\pi)z}{\cos z} + \lim_{z \to 2\pi}\frac{z^2+2(z-2\pi)z}{\cos z} \right]$$

$$\Rightarrow I = 2\pi i \left[\frac{0}{\cos 0} + \frac{(-\pi)^2+2(-\pi+\pi)(-\pi)}{\cos(-\pi)} + \frac{(\pi)^2+2(\pi-\pi)(\pi)}{\cos(\pi)} + \frac{(-2\pi)^2+2(-2\pi+2\pi)(-2\pi)}{\cos(-2\pi)} + \frac{(2\pi)^2+2(2\pi-2\pi)(2\pi)}{\cos(2\pi)} \right]$$

$$\Rightarrow I = 2\pi i \left[0 + \frac{\pi^2}{\cos(-\pi)} + \frac{\pi^2}{\cos(\pi)} + \frac{4\pi^2}{\cos(-2\pi)} + \frac{4\pi^2}{\cos(2\pi)} \right]$$

$$\Rightarrow I = 2\pi i \left[0 - \pi^2 - \pi^2 + 4\pi^2 + 4\pi^2 \right]$$

$$\Rightarrow I = 12\pi^3 i$$

Choice (4) is the answer.

Notes

In this problem, the relations below have been used:

The singular points of the complex function $f(z) = \frac{p(z)}{q(z)}$ can be determined by finding the roots of the equation $q(z) = 0$.

For a circle with equation $|z - z_0| = R$, the point z_1 is inside the circle, on the circle, and outside the circle, respectively, if:

$$|z_1 - z_0| < R$$

$$|z_1 - z_0| = R$$

$$|z_1 - z_0| > R$$

The residue of the function $f(z)$ at the given singularity can be calculated by using two methods as follows:

Method 1: The coefficient of $\frac{1}{z - z_0}$ in the Taylor series expansion of the function $f(z)$ at the essential singular point $z = z_0$ as well as the coefficient of $\frac{1}{z}$ in the Laurent series expansion of the function $f(z)$ at the essential singular point $z = 0$ is called residue.

Method 2: If z_0 is the mth-order pole of a complex function $f(z)$, then the residue of the function at $z = z_0$ can be calculated as follows:

$$\operatorname*{Res}_{z=z_0} f(z) = \frac{1}{(m-1)!} \lim_{z \to z_0} \frac{d^{m-1}}{dz^{m-1}} \left[(z - z_0)^m f(z) \right]$$

If z_0 is the first-order pole of a complex function $f(z)$, then the residue of the function at $z = z_0$ can be calculated as follows:

$$\operatorname*{Res}_{z=z_0} f(z) = \lim_{z \to z_0} (z - z_0) f(z)$$

$$h(z) = \lim_{z \to z_0} \frac{f(z)}{g(z)} = \frac{0}{0} \overset{H}{\Rightarrow} h(z) = \lim_{z \to z_0} \frac{f'(z)}{g'(z)}$$

$$\frac{d}{dz} (\sin z) = \cos z$$

$$\cos 0 = 1$$

$$\cos(-\pi) = \cos(\pi) = -1$$

$$\cos(-2\pi) = \cos(2\pi) = 1$$

8.17. Based on the information given in the problem, we have:

$$I = \oint_C \frac{z^3 + 3}{z(z - i)^2}$$

In addition, the contour C is shown in Fig. 8.1.

If $f(z)$ is holomorphic on and inside the counterclockwise contour C (a closed loop that does not intersect itself) except at the finite number of singular points z_1, z_2, ..., z_n and the contour surrounds these singular points, then the contour integral can be calculated as follows:

$$I = \oint_C f(z) dz = 2\pi i \left(\operatorname*{Res}_{z=z_1} f(z) + \operatorname*{Res}_{z=z_2} f(z) + \ldots + \operatorname*{Res}_{z=z_n} f(z) \right)$$

Moreover, if the direction of the contour C is clockwise, the contour integral is calculated as follows:

$$I = \oint_C f(z)dz = -2\pi i \left(\operatorname*{Res}_{z=z_1} f(z) + \operatorname*{Res}_{z=z_2} f(z) + \ldots + \operatorname*{Res}_{z=z_n} f(z) \right)$$

The singular point of the complex function can be calculated as follows:

$$z(z-i)^2 \Rightarrow z = 0, i$$

The singular point $z = i$ is a second-order pole. Moreover, as can be noticed from Fig. 8.1, the contour around the singular point $z = 0$ and $z = i$ are clockwise and counterclockwise, respectively. Therefore:

$$I = 2\pi i \left[-\operatorname*{Res}_{z=0} f(z) + \operatorname*{Res}_{z=i} f(z) \right]$$

$$\Rightarrow I = 2\pi i \left[-\lim_{z \to 0} (z-0)\frac{z^3+3}{z(z-i)^2} + \lim_{z \to i}\frac{d}{dz}\left[(z-i)^2 \frac{z^3+3}{z(z-i)^2} \right] \right]$$

$$\Rightarrow I = 2\pi i \left[-\lim_{z \to 0}\frac{z^3+3}{(z-i)^2} + \lim_{z \to i}\frac{d}{dz}\left[\frac{z^3+3}{z} \right] \right]$$

$$\Rightarrow I = 2\pi i \left[-\lim_{z \to 0}\frac{z^3+3}{(z-i)^2} + \lim_{z \to i}\frac{3z^2 z - (z^3+3)}{z^2} \right]$$

$$\Rightarrow I = 2\pi i \left[-\frac{3}{i^2} + \frac{2i^3-3}{i^2} \right] = 2\pi i\left[3 + \frac{-2i-3}{-1} \right] = 2\pi i[6+2i]$$

$$\Rightarrow I = 4\pi(-1+3i)$$

Choice (1) is the answer.

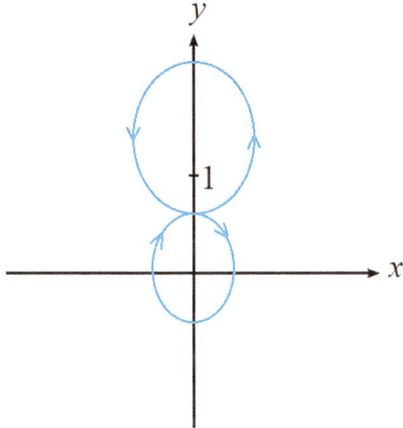

Fig. 8.1 The contour related to Problem 8.17

Notes

In this problem, the relations below have been used:

The singular points of the complex function $f(z) = \frac{p(z)}{q(z)}$ can be determined by finding the roots of the equation $q(z) = 0$.

The residue of the function $f(z)$ at the given singularity can be calculated by using two methods as follows:

Method 1: The coefficient of $\frac{1}{z-z_0}$ in the Taylor series expansion of the function $f(z)$ at the essential singular point $z = z_0$ as well as the coefficient of $\frac{1}{z}$ in the Laurent series expansion of the function $f(z)$ at the essential singular point $z = 0$ is called residue.

Method 2: If z_0 is the mth-order pole of a complex function $f(z)$, then the residue of the function at $z = z_0$ can be calculated as follows:

$$\operatorname*{Res}_{z=z_0} f(z) = \frac{1}{(m-1)!} \lim_{z \to z_0} \frac{d^{m-1}}{dz^{m-1}} \left[(z-z_0)^m f(z) \right]$$

If z_0 is the first-order pole of a complex function $f(z)$, then the residue of the function at $z = z_0$ can be calculated as follows:

$$\operatorname*{Res}_{z=z_0} f(z) = \lim_{z \to z_0} (z - z_0) f(z)$$

$$i^2 = -1$$

$$\frac{d}{dz} \left(\frac{f(z)}{g(z)} \right) = \frac{f'(z)g(z) - g'(z)f(z)}{(g(z))^2}$$

8.18. Based on the information given in the problem, we know that the direction of contour is counterclockwise. In addition:

$$I = \frac{3}{\pi i} \oint_{|z|=3} \frac{z^5}{(z-2)^4} dz$$

If $f(z)$ is holomorphic on and inside the counterclockwise contour C (a closed loop that does not intersect itself) except at the finite number of singular points $z_1, z_2, \ldots, z_n$ and the contour surrounds these singular points, then the contour integral can be calculated as follows:

$$I = \oint_C f(z)dz = 2\pi i \left(\operatorname*{Res}_{z=z_1} f(z) + \operatorname*{Res}_{z=z_2} f(z) + \ldots + \operatorname*{Res}_{z=z_n} f(z) \right)$$

Moreover, if the direction of the contour C is clockwise, the contour integral is calculated as follows:

$$I = \oint_C f(z)dz = -2\pi i \left(\operatorname*{Res}_{z=z_1} f(z) + \operatorname*{Res}_{z=z_2} f(z) + \ldots + \operatorname*{Res}_{z=z_n} f(z) \right)$$

The singular point of the complex function can be calculated as follows:

$$(z-2)^4 \Rightarrow z = 2$$

The singular point $z = 2$, which is a four-order pole, is inside the contour $|z| = 3$, as can be seen in the following:

$$z = 2 \Rightarrow |2| = 2 < 3$$

Therefore:

$$I = \frac{3}{\pi i} \times 2\pi i \left[\operatorname*{Res}_{z=2} f(z) \right]$$

$$\Rightarrow I = 6 \left[\frac{1}{3!} \lim_{z \to 2} \frac{d^3}{dz^3} \left[(z-2)^4 \frac{z^5}{(z-2)^4} \right] \right]$$

$$\Rightarrow I = \lim_{z \to 2} \frac{d^3}{dz^3} \left(z^5 \right) = \lim_{z \to 2} \frac{d^2}{dz^2} \left(5z^4 \right) = \lim_{z \to 2} \frac{d}{dz} \left(20z^3 \right) = \lim_{z \to 2} 60z^2$$

$$\Rightarrow I = 60 \times 2^2$$

$$\Rightarrow I = 240$$

Choice (2) is the answer.

Notes

In this problem, the relations below have been used:

The singular points of the complex function $f(z) = \frac{p(z)}{q(z)}$ can be determined by finding the roots of the equation $q(z) = 0$.

The residue of the function $f(z)$ at the given singularity can be calculated by using two methods as follows:

Method 1: The coefficient of $\frac{1}{z - z_0}$ in the Taylor series expansion of the function $f(z)$ at the essential singular point $z = z_0$ as well as the coefficient of $\frac{1}{z}$ in the Laurent series expansion of the function $f(z)$ at the essential singular point $z = 0$ is called residue.

Method 2: If z_0 is the mth-order pole of a complex function $f(z)$, then the residue of the function at $z = z_0$ can be calculated as follows:

$$\operatorname*{Res}_{z=z_0} f(z) = \frac{1}{(m-1)!} \lim_{z \to z_0} \frac{d^{m-1}}{dz^{m-1}} \left[(z - z_0)^m f(z) \right]$$

If z_0 is the first-order pole of a complex function $f(z)$, then the residue of the function at $z = z_0$ can be calculated as follows:

$$\operatorname*{Res}_{z=z_0} f(z) = \lim_{z \to z_0} (z - z_0) f(z)$$

$$\frac{d}{dz} z^n = n z^{n-1}$$

References

1. Rahmani-Andebili, M. (2024). Precalculus (2nd Ed.) – Practice Problems, Methods, and Solutions, Springer Nature.
2. Rahmani-Andebili, M. (2023). Calculus III – Practice Problems, Methods, and Solutions, Springer Nature.
3. Rahmani-Andebili, M. (2023). Calculus II – Practice Problems, Methods, and Solutions, Springer Nature.
4. Rahmani-Andebili, M. (2023). Calculus I (2nd Ed.) – Practice Problems, Methods, and Solutions, Springer Nature.
5. Rahmani-Andebili, M. (2022). Differential Equations – Practice Problems, Methods, and Solutions, Springer Nature.
6. Rahmani-Andebili, M. (2021). Calculus – Practice Problems, Methods, and Solutions, Springer Nature.
7. Rahmani-Andebili, M. (2021). Precalculus – Practice Problems, Methods, and Solutions, Springer Nature.

Fourier Series, Half-Domain Fourier Sine and Cosine Series, Complex Fourier Series, Fourier Integral, Complex Fourier Integral, Fourier Transform, and Half-Domain Fourier Sine and Cosine Transforms: Problems

Abstract

In this chapter, the basic and advanced problems concerned with the Fourier series of periodic functions, half-domain Fourier sine and cosine series of aperiodic functions, complex Fourier series of periodic functions, Fourier integral of aperiodic functions, complex Fourier integral of aperiodic functions, Fourier transform of aperiodic functions, and half-domain Fourier sine and cosine transforms of aperiodic functions are presented and studied. Herein, different types of problems and exercises are presented that are categorized as follows:

○ *Problems with detailed solution*: They have been designed to teach students the subjects in detail. Moreover, they have been categorized into different levels based on their difficulty levels (easy, normal, and hard) and calculation amounts (small, normal, and large).

○ *Partially solved exercises*: They have been designed to encourage students to practice more problems while guiding them through the problem-solving procedure and hinting the required formulas.

○ *Exercises with final answer*: They have been designed to encourage students to practice by themselves while hinting them by the final answer as well as to help instructors to give tests or quizzes.

9.1   Fourier Series of Periodic Functions

9.1. Calculate the constant value of the Fourier series of the periodic function $f(x) = x^2$, $-\pi \leq x < \pi$ [1–7].

Difficulty level   ● Easy   ○ Normal   ○ Hard
Calculation amount   ● Small   ○ Normal   ○ Large

1) $a_0 = \dfrac{3\pi}{2}$

2) $a_0 = \dfrac{\pi^2}{3}$

3) $a_0 = \dfrac{2\pi^2}{3}$

4) $a_0 = 3\pi^3$

Exercise

Calculate the constant value of the Fourier series of the periodic function $f(x) = x^3$, $-\pi \leq x < \pi$.

Final Answer

0.

9.2. Calculate the coefficient of sine terms in the Fourier series of the periodic function below:

$$f(x) = \begin{cases} -k & -\pi \le x < 0 \\ k & 0 \le x < \pi \end{cases}$$

Difficulty level   ○ Easy   ● Normal   ○ Hard
Calculation amount   ○ Small   ● Normal   ○ Large
1) $b_n = 0$
2) $b_n = \frac{2k}{n\pi}(1 + \cos n\pi)$
3) $b_n = \frac{2k}{n\pi}(1 - \cos n\pi)$
4) $b_n = \frac{2k}{n\pi}(1 - \sin n\pi)$

9.3. Calculate the Fourier series of the periodic function below:

$$f(x) = \begin{cases} 1 & 0 \le x < 1 \\ 0 & 1 \le x < 2 \end{cases}$$

Difficulty level   ○ Easy   ● Normal   ○ Hard
Calculation amount   ○ Small   ○ Normal   ● Large
1) $f(x) = \frac{1}{2} + \sum_{k=1}^{\infty} \frac{2}{(2k-1)\pi} \sin(2k-1)\pi x$
2) $f(x) = \frac{1}{2} + \sum_{k=1}^{\infty} \frac{2}{(2k-1)\pi} \cos(2k-1)\pi x$
3) $f(x) = \sum_{k=1}^{\infty} \frac{2}{(2k-1)\pi} \cos(2k-1)\pi x$
4) $f(x) = \sum_{k=1}^{\infty} \frac{2}{(2k-1)\pi} \sin(2k-1)\pi x$

Partially Solved Exercise

Calculate the Fourier series of the following periodic function:

$$f(x) = \begin{cases} -c & -\pi \le x < 0 \\ c & 0 \le x < \pi \end{cases}$$

Solution

As can be noticed, the function is odd since $f(-x) = -f(x)$.

The Fourier series of a periodic odd function can be calculated as follows:

$$a_0 = 0$$

$$a_n = 0$$

$$b_n = \frac{4}{T} \int_0^{\frac{T}{2}} f(x) \sin\left(\frac{2n\pi}{T}x\right) dx$$

$$f(x) = \sum_{n=1}^{\infty} b_n \sin\left(\frac{2n\pi}{T}x\right)$$

Therefore:

$$b_n = \frac{4}{2\pi} \int_0^{\pi} c \sin nx \; dx$$

$$\Rightarrow b_n = (\qquad)[\qquad\qquad]_0^{\pi}$$

$$\Rightarrow b_n = (\qquad)[\qquad\quad]$$

$$\Rightarrow f(x) = (\qquad) \sum_{n=1}^{\infty} [\qquad\qquad]$$

$$\Rightarrow f(x) = \frac{4c}{\pi} \sum_{n=1}^{\infty} \frac{\sin(2n-1)x}{(2n-1)}$$

Notes

In this problem, the relations below have been used:

$$\int \sin ax \; dx = -\frac{1}{a} \cos ax$$

$$\cos 0 = 1$$

$$\cos n\pi = (-1)^n$$

$$1 - (-1)^k = \begin{cases} 2 & k = 2n-1 \\ 0 & k = 2n \end{cases}$$

9.4. The Fourier series of the periodic function $f(x) = 4 - x^2$, $-2 \leq x < 2$ is as follows:

$$f(x) = \frac{8}{3} + \frac{16}{\pi^2}\left(\cos\frac{\pi x}{2} - \frac{1}{2^2}\cos\frac{2\pi x}{2} + \frac{1}{3^2}\cos\frac{3\pi x}{2} - \cdots\right)$$

Calculate the value of the following term:

$$S = 1 + \frac{1}{2^2} + \frac{1}{3^2} + \cdots$$

Difficulty level   ○ Easy   ○ Normal   ● Hard
Calculation amount   ○ Small   ● Normal   ○ Large

1) $\dfrac{\pi^2}{4}$

2) $\dfrac{\pi^2}{6}$

3) $\dfrac{\pi^2}{8}$

4) $\dfrac{\pi^2}{12}$

9.5. In the Fourier series of the periodic function below, calculate the value of b_1:

$$f(x) = \begin{cases} \sin x & -\pi \leq x < 0 \\ 0 & 0 \leq x < \pi \end{cases}$$

Difficulty level   ○ Easy   ● Normal   ○ Hard
Calculation amount   ○ Small   ● Normal   ○ Large

1) $-\dfrac{1}{2}$

2) -2

3) $\dfrac{1}{2}$

4) 2

9.6. In the Fourier series of the periodic function below, calculate the coefficient of $\sin 3x$:

$$f(x) = \begin{cases} \dfrac{-1}{2}\pi & -\pi \leq x < 0 \\ \dfrac{1}{2}\pi & 0 \leq x < \pi \end{cases}$$

Difficulty level   ○ Easy   ● Normal   ○ Hard
Calculation amount   ○ Small   ● Normal   ○ Large

1) $\dfrac{-2}{3}$

2) $\dfrac{-1}{3}$

3) $\dfrac{2}{3}$

4) $\dfrac{1}{3}$

9.2   Half-Domain Fourier Sine and Cosine Series of Aperiodic Functions

9.7. Calculate the half-domain Fourier cosine series of the following aperiodic function:

$$f(x) = \begin{cases} 1 & 0 \leq x < \pi \\ 0 & \pi \leq x < 2\pi \end{cases}$$

Difficulty level   ○ Easy   ● Normal   ○ Hard
Calculation amount   ○ Small   ● Normal   ○ Large

1) $\dfrac{1}{2} + \displaystyle\sum_{n=1}^{\infty} \dfrac{2}{n\pi} \sin\left(\dfrac{n\pi}{2}\right) \cos\left(\dfrac{nx}{2}\right)$

2) $\dfrac{1}{2} + \displaystyle\sum_{n=1}^{\infty} \dfrac{1}{n\pi} \sin\left(\dfrac{n\pi}{2}\right) \cos\left(\dfrac{nx}{2}\right)$

3) $\dfrac{1}{2} + \displaystyle\sum_{n=1}^{\infty} \dfrac{4}{n\pi} \sin\left(\dfrac{n\pi}{2}\right) \cos\left(\dfrac{nx}{2}\right)$

4) $\displaystyle\sum_{n=1}^{\infty} \dfrac{2}{n\pi} \sin\left(\dfrac{n\pi}{2}\right) \cos\left(\dfrac{nx}{2}\right)$

9.8. Calculate the half-domain Fourier sine series of the following aperiodic function:

$$f(x) = \begin{cases} 1 & 0 \le x < \pi \\ 0 & \pi \le x < 2\pi \end{cases}$$

Difficulty level   ○ Easy   ● Normal   ○ Hard
Calculation amount   ○ Small   ● Normal   ○ Large

1) $\displaystyle\sum_{n=1}^{\infty} \left(-\dfrac{2}{n\pi}\left[\cos\left(\dfrac{n\pi}{2}\right) - 1\right]\right) \sin\left(\dfrac{nx}{2}\right)$

2) $\displaystyle\sum_{n=1}^{\infty} -\dfrac{2}{n\pi} \cos\left(\dfrac{n\pi}{2}\right) \sin\left(\dfrac{nx}{2}\right)$

3) $\displaystyle\sum_{n=1}^{\infty} \left(\dfrac{2}{n\pi}\left[\cos\left(\dfrac{n\pi}{2}\right) - 1\right]\right) \sin\left(\dfrac{nx}{2}\right)$

4) $\displaystyle\sum_{n=1}^{\infty} \left(\dfrac{2}{n\pi}\left[\cos\left(\dfrac{n\pi}{2}\right) + 1\right]\right) \sin\left(\dfrac{nx}{2}\right)$

9.9. The Fourier series of the aperiodic function $f(x) = |\sin x|$, $-\pi \le x < \pi$ is as follows:

$$f(x) = \dfrac{2}{\pi} + \dfrac{4}{\pi} \sum_{n=1}^{\infty} \dfrac{\cos 2nx}{1 - 4n^2}$$

Calculate the Fourier sine series of the aperiodic function $g(x) = \cos x$, $0 \le x < \pi$.
Difficulty level   ○ Easy   ● Normal   ○ Hard
Calculation amount   ○ Small   ● Normal   ○ Large

1) $-\dfrac{8}{\pi} \displaystyle\sum_{n=1}^{\infty} \dfrac{\sin 2nx}{1 - 4n^2}$

2) $-\dfrac{8}{\pi} \displaystyle\sum_{n=1}^{\infty} \dfrac{n \sin 2nx}{1 - 4n^2}$

3) $\dfrac{4}{\pi} \displaystyle\sum_{n=1}^{\infty} \dfrac{\sin 2nx}{1 - 4n^2}$

4) $\dfrac{4}{\pi} \displaystyle\sum_{n=1}^{\infty} \dfrac{n \sin 2nx}{1 - 4n^2}$

9.10. The Fourier series of the aperiodic function $f(x) = \begin{cases} -1 & -\pi \le x < 0 \\ 1 & 0 \le x < \pi \end{cases}$ is as follows:

$$f(x) = \dfrac{4}{\pi}\left[\dfrac{\sin x}{1} + \dfrac{\sin 3x}{3} + \dfrac{\sin 5x}{5} + \cdots\right]$$

Calculate the Fourier sine series of the aperiodic function $g(x) = |x|,\ -\pi \le x < \pi$.

1) $\dfrac{4}{\pi}\displaystyle\sum_{k=0}^{\infty}\left[\dfrac{\cos(2k+1)x}{(2k+1)^2}\right]$

2) $\dfrac{4}{\pi}\displaystyle\sum_{k=0}^{\infty}\left[\dfrac{\sin(2k+1)x}{(2k+1)^2}\right]$

3) $\dfrac{4}{\pi}\displaystyle\sum_{k=0}^{\infty}\left[\dfrac{1-\cos(2k+1)x}{(2k+1)^2}\right]$

4) $\dfrac{4}{\pi}\displaystyle\sum_{k=0}^{\infty}\left[\dfrac{1-\sin(2k+1)x}{(2k+1)^2}\right]$

9.3   Complex Fourier Series of Periodic Functions

9.11. Calculate the complex Fourier series of the periodic function $f(x) = x^2,\ 0 \le x < 2\pi$.

1) $f(x) = \dfrac{4\pi^2}{3} + \displaystyle\sum_{n=-\infty}^{\infty}\dfrac{2+i2\pi n}{n^2}e^{-inx}$

2) $f(x) = \dfrac{2\pi^2}{3} + \displaystyle\sum_{n=-\infty}^{\infty}\dfrac{2+i2\pi n}{n^2}e^{inx}$

3) $f(x) = \dfrac{2\pi^2}{3} + \displaystyle\sum_{n=-\infty}^{\infty}\dfrac{2+i2\pi n}{n^2}e^{-inx}$

4) $f(x) = \dfrac{4\pi^2}{3} + \displaystyle\sum_{n=-\infty}^{\infty}\dfrac{2+i2\pi n}{n^2}e^{inx}$

9.4   Fourier Integral of Aperiodic Functions

9.12. Calculate the Fourier integral of the function below:

$$f(x) = \begin{cases} \sin x & 0 \le x < \pi \\ 0 & \text{Otherwise} \end{cases}$$

1) $\dfrac{1}{\pi}\displaystyle\int_0^{\infty}\left[\dfrac{1+\cos\omega\pi}{1-\omega^2}\cos\omega x - \dfrac{\sin\omega\pi}{1-\omega^2}\sin\omega x\right]d\omega$

2) $\dfrac{1}{2\pi}\displaystyle\int_0^{\infty}\dfrac{\cos\omega x + \cos\omega(\pi-x)}{1-\omega^2}d\omega$

3) $\dfrac{1}{\pi}\displaystyle\int_0^{\infty}\dfrac{\cos\omega x + \cos(\pi-x)}{1-\omega^2}d\omega$

4) $\dfrac{1}{2\pi}\displaystyle\int_0^{\infty}\dfrac{\cos\omega x + \cos(\pi-x)}{1-\omega^2}d\omega$

9.13. Calculate the value of the following integral by using the Fourier integral of the function $f(x) = e^{-x}$, $x > 0$:

$$I = \int_0^\infty \frac{\cos \omega x + \omega \sin \omega x}{1 + \omega^2} d\omega$$

Difficulty level   ○ Easy   ○ Normal   ● Hard
Calculation amount   ○ Small   ● Normal   ○ Large

1) 0

2) $\dfrac{\pi}{2}$

3) πe^{-x}

4) $\dfrac{\pi}{2} e^{-x}$

9.5   Complex Fourier Integral of Aperiodic Functions

9.14. Calculate the complex Fourier integral of the function below:

$$f(x) = \begin{cases} \sinh 2x & 0 \le x < 2 \\ 0 & x \ge 2 \end{cases}$$

Difficulty level   ○ Easy   ○ Normal   ● Hard
Calculation amount   ○ Small   ○ Normal   ● Large

1) $\dfrac{1}{2\pi} \displaystyle\int_{-\infty}^{+\infty} \left[\dfrac{e^{-i2\omega}(4\cosh 4 + i2\omega \sinh 4)}{4 + \omega^2} \right] e^{i\omega x} d\omega$

2) $\dfrac{1}{4\pi} \displaystyle\int_{-\infty}^{+\infty} \left[\dfrac{e^{-i2\omega}(4\cosh 4 + i\omega \sinh 4) - 4}{4 + \omega^2} \right] e^{i\omega x} d\omega$

3) $\dfrac{1}{4\pi} \displaystyle\int_{-\infty}^{+\infty} \left[\dfrac{e^{-i2\omega}(2\cosh 4 + i2\omega \sinh 4) - 4}{4 + \omega^2} \right] e^{i\omega x} d\omega$

4) $\dfrac{1}{4\pi} \displaystyle\int_{-\infty}^{+\infty} \left[\dfrac{e^{-i2\omega}(4\cosh 4 + i2\omega \sinh 4) - 4}{4 + \omega^2} \right] e^{i\omega x} d\omega$

9.6   Fourier Transform of Aperiodic Functions

9.15. Calculate the Fourier transform of the aperiodic function below:

$$f(x) = \begin{cases} 1 & |x| \le a \\ 0 & |x| > a \end{cases}$$

Difficulty level   ● Easy   ○ Normal   ○ Hard
Calculation amount   ● Small   ○ Normal   ○ Large

1) $\dfrac{2 \sin \omega a}{\omega}$

2) $\dfrac{2 \cos \omega a}{\omega}$

3) $\cos \omega a$

4) $\sin \omega a$

9.16. Calculate the Fourier transform of the solution $(Y(\omega))$ of the following second-order differential equation:

$$y'' + 4y = \begin{cases} 0 & x \le 0 \\ e^{-2x} & x > 0 \end{cases}$$

Difficulty level   ○ Easy   ○ Normal   ● Hard
Calculation amount   ● Small   ○ Normal   ○ Large

1) $Y(\omega) = \dfrac{1}{(i\omega + 2)(4 + \omega^2)}$

2) $Y(\omega) = \dfrac{1}{(i\omega + 2)(4 - \omega^2)}$

3) $Y(\omega) = \dfrac{1}{(2 - i\omega)(4 + \omega^2)}$

4) $Y(\omega) = \dfrac{1}{(2 - i\omega)(4 - \omega^2)}$

9.17. Calculate the Fourier transform of the solution $(Y(\omega))$ of the following first-order differential equation:

$$y - 4y = \begin{cases} e^{-4x} & x \ge 0 \\ 0 & x < 0 \end{cases}$$

Difficulty level   ○ Easy   ○ Normal   ● Hard
Calculation amount   ● Small   ○ Normal   ○ Large

1) $\dfrac{-1}{16 + \omega^2}$

2) $\dfrac{1}{4 - j\omega}$

3) $\dfrac{1}{4 + j\omega}$

4) $\dfrac{1}{16 - \omega^2}$

9.7   Half-Domain Fourier Sine and Cosine Transforms of Aperiodic Functions

9.18. The Fourier cosine series of the aperiodic function $f(x) = x,\ 0 \le x < \pi$ is as follows:

$$f(x) = \frac{\pi}{2} - \frac{4}{\pi}\left(\cos x + \frac{\cos 3x}{3^2} + \frac{\cos 5x}{5^2} + \cdots\right)$$

Calculate the Fourier sine series of the aperiodic function $g(x) = x(\pi - x)\frac{\pi}{8}, 0 \le x < \pi$.

Difficulty level   ○ Easy   ○ Normal   ● Hard
Calculation amount   ○ Small   ● Normal   ○ Large

1) $\displaystyle\sum_{n=1}^{\infty} \frac{\sin 2nx}{(2n)^3}$

2) $\displaystyle\sum_{n=1}^{\infty} \frac{\sin(2n+1)x}{(2n+1)^3}$

3) $\displaystyle\sum_{n=1}^{\infty} \frac{\sin 2nx}{n^3}$

4) $\displaystyle\sum_{n=1}^{\infty} \frac{\sin(2n-1)x}{(2n-1)^3}$

References

1. Rahmani-Andebili, M. (2024). Precalculus (2nd Ed.) – Practice Problems, Methods, and Solutions, Springer Nature.
2. Rahmani-Andebili, M. (2023). Calculus III – Practice Problems, Methods, and Solutions, Springer Nature.
3. Rahmani-Andebili, M. (2023). Calculus II – Practice Problems, Methods, and Solutions, Springer Nature.
4. Rahmani-Andebili, M. (2023). Calculus I (2nd Ed.) – Practice Problems, Methods, and Solutions, Springer Nature.
5. Rahmani-Andebili, M. (2022). Differential Equations – Practice Problems, Methods, and Solutions, Springer Nature.
6. Rahmani-Andebili, M. (2021). Calculus – Practice Problems, Methods, and Solutions, Springer Nature.
7. Rahmani-Andebili, M. (2021). Precalculus – Practice Problems, Methods, and Solutions, Springer Nature.

Fourier Series, Half-Domain Fourier Sine and Cosine Series, Complex Fourier Series, Fourier Integral, Complex Fourier Integral, Fourier Transform, and Half-Domain Fourier Sine and Cosine Transforms: Solutions of Problems

10

Abstract

In this chapter, the problems of the ninth chapter are fully solved, in detail, step-by-step, and with different methods.

10.1 Fourier Series of Periodic Functions

10.1. Based on the information given in the problem, we have [1–7]:

$$f(x) = x^2, \quad -\pi \leq x < \pi$$

The constant value (DC value) of the Fourier series of a periodic function can be calculated as follows:

$$a_0 = \frac{1}{T} \int_{-\frac{T}{2}}^{\frac{T}{2}} f(x) dx$$

Therefore:

$$a_0 = \frac{1}{2\pi} \int_{-\pi}^{\pi} x^2 dx = \frac{1}{2\pi} \left[\frac{x^3}{3} \right]_{-\pi}^{\pi}$$

$$\Rightarrow a_0 = \frac{1}{2\pi} \left(\frac{\pi^3}{3} - \frac{(-\pi)^3}{3} \right)$$

$$\Rightarrow a_0 = \frac{\pi^2}{3}$$

Choice (2) is the answer.

Notes

In this problem, the relation below has been used:

$$\int x^n dx = \frac{1}{n+1} x^{n+1}$$

10.2. Based on the information given in the problem, we have:

$$f(x) = \begin{cases} -k & -\pi \leq x < 0 \\ k & 0 \leq x < \pi \end{cases},$$

The coefficient of sine terms in the Fourier series of a periodic function can be calculated as follows:

$$b_n = \frac{2}{T} \int_{-\frac{T}{2}}^{\frac{T}{2}} f(x) \sin nx \, dx$$

Therefore:

$$b_n = \frac{2}{2\pi} \left[\int_{-\pi}^{0} -k \sin nx \, dx + \int_{0}^{\pi} k \sin nx \, dx \right] = \frac{k}{n\pi} [\cos nx]_{-\pi}^{0} - \frac{k}{n\pi} [\cos nx]_{0}^{\pi}$$

$$\Rightarrow b_n = \frac{k}{n\pi} (\cos 0 - \cos(-n\pi)) - \frac{k}{n\pi} (\cos(n\pi) - \cos 0)$$

$$\Rightarrow b_n = \frac{k}{n\pi} (1 - \cos(n\pi)) - \frac{k}{n\pi} (\cos(n\pi) - 1)$$

$$\Rightarrow b_n = \frac{2k}{n\pi} (1 - \cos n\pi)$$

Choice (3) is the answer.

Notes

In this problem, the relations below have been used:

$$\int \sin ax \, dx = -\frac{1}{a} \cos ax$$

$$\cos 0 = 1$$

10.3. Based on the information given in the problem, we have:

$$f(x) = \begin{cases} 1 & 0 \leq x < 1 \\ 0 & 1 \leq x < 2 \end{cases}$$

The Fourier series of a periodic function can be calculated as follows:

$$f(x) = a_0 + \sum_{n=1}^{\infty} \left(a_n \cos\left(\frac{2n\pi}{T}x\right) + b_n \sin\left(\frac{2n\pi}{T}x\right) \right)$$

where

$$a_0 = \frac{1}{T} \int_{-\frac{T}{2}}^{\frac{T}{2}} f(x) dx$$

$$a_n = \frac{2}{T} \int_{-\frac{T}{2}}^{\frac{T}{2}} f(x) \cos\left(\frac{2n\pi}{T} x\right) dx$$

$$b_n = \frac{2}{T} \int_{-\frac{T}{2}}^{\frac{T}{2}} f(x) \sin\left(\frac{2n\pi}{T} x\right) dx$$

Therefore:

$$a_0 = \frac{1}{2} \int_0^1 dx + \frac{1}{2} \int_1^2 0 dx = \frac{1}{2} [x]_0^1$$

$$\Rightarrow a_0 = \frac{1}{2}$$

$$a_n = \frac{2}{2} \int_0^1 \cos n\pi x dx + \frac{2}{2} \int_1^2 0 \times \cos n\pi x dx = \frac{1}{n\pi} [\sin n\pi x]_0^1 = \frac{1}{n\pi} (\sin n\pi - \sin 0)$$

$$\Rightarrow a_n = 0$$

$$b_n = \frac{2}{2} \int_0^1 \sin n\pi x dx + \frac{2}{2} \int_1^2 0 \times \sin n\pi x dx = \frac{1}{n\pi} [-\cos n\pi x]_0^1 = \frac{1}{n\pi} (\cos 0 - \cos n\pi)$$

$$\Rightarrow b_n = \frac{1 - (-1)^n}{n\pi}$$

$$f(x) = \frac{1}{2} + \sum_{n=1}^{\infty} \left(0 \times \cos(n\pi x) + \frac{1 - (-1)^n}{n\pi} \sin(n\pi x) \right) = \frac{1}{2} + \sum_{n=1}^{\infty} \frac{1 - (-1)^n}{n\pi} \sin(n\pi x)$$

$$\Rightarrow f(x) = \frac{1}{2} + \sum_{n=1}^{\infty} \frac{2}{(2n-1)\pi} \sin(2n-1)\pi x$$

Choice (1) is the answer.

Notes

In this problem, the relations below have been used:

$$\int x^n dx = \frac{1}{n+1} x^{n+1}$$

$$\int \cos ax \, dx = \frac{1}{a} \sin ax$$

$$\int \sin ax \, dx = -\frac{1}{a}\cos ax$$

$$\sin n\pi = 0$$

$$\sin 0 = 0$$

$$\cos 0 = 1$$

$$\cos n\pi = (-1)^n$$

$$1-(-1)^k = \begin{cases} 2 & k = 2n-1 \\ 0 & k = 2n \end{cases}$$

10.4. Based on the information given in the problem, we need to calculate the value of the term $S = 1 + \frac{1}{2^2} + \frac{1}{3^2} + \cdots$ by using the periodic function below:

$$f(x) = 4 - x^2, \quad -2 \leq x < 2 \tag{1}$$

Moreover, the Fourier series of the function is as follows:

$$f(x) = \frac{8}{3} + \frac{16}{\pi^2}\left(\cos\frac{\pi x}{2} - \frac{1}{2^2}\cos\frac{2\pi x}{2} + \frac{1}{3^2}\cos\frac{3\pi x}{2} - \cdots\right) \tag{2}$$

The value of function at $x = 2$ can be calculated by using (1) and (2) as follows:

$$f(2) = 4 - 4 = 0 \tag{3}$$

$$f(2) = \frac{8}{3} + \frac{16}{\pi^2}\left[\cos\frac{\pi \times 2}{2} - \frac{1}{2^2}\cos\frac{2\pi \times 2}{2} + \frac{1}{3^2}\cos\frac{3\pi \times 2}{2} - \cdots\right]$$

$$\Rightarrow f(2) = \frac{8}{3} + \frac{16}{\pi^2}\left[-1 - \frac{1}{2^2} - \frac{1}{3^2} - \frac{1}{4^2} - \cdots\right] \tag{4}$$

Solving (3) and (4):

$$0 = \frac{8}{3} + \frac{16}{\pi^2}\left(-1 - \frac{1}{2^2} - \frac{1}{3^2} - \frac{1}{4^2} - \cdots\right)$$

$$\Rightarrow \frac{16}{\pi^2}\left(-1 - \frac{1}{2^2} - \frac{1}{3^2} - \frac{1}{4^2} - \cdots\right) = -\frac{8}{3}$$

$$\Rightarrow 1 + \frac{1}{2^2} + \frac{1}{3^2} + \frac{1}{4^2} + \cdots = \frac{\pi^2}{6}$$

Choice (2) is the answer.

Notes

In this problem, the relations below have been used:

$$\cos \pi = -1$$

$$\cos 2\pi = 1$$

10.5. Based on the information given in the problem, we have:

$$f(x) = \begin{cases} \sin x & -\pi \leq x < 0 \\ 0 & 0 \leq x < \pi \end{cases},$$

As we know, the coefficient of sine terms in the Fourier series of a periodic function can be calculated as follows:

$$b_n = \frac{2}{T} \int_{-\frac{T}{2}}^{\frac{T}{2}} f(x) \sin nx \, dx$$

Therefore:

$$b_1 = \frac{2}{2\pi} \int_{-\pi}^{0} \sin^2 x \, dx$$

$$\Rightarrow b_1 = \frac{1}{\pi} \int_{-\pi}^{0} \left[\frac{1 - \cos 2x}{2} \right] dx = \frac{1}{2\pi} \left[\int_{-\pi}^{0} dx - \int_{-\pi}^{0} \cos 2x \, dx \right]$$

$$\Rightarrow b_1 = \frac{1}{2\pi} \left([x]_{-\pi}^{0} - \left[\frac{\sin 2x}{2} \right]_{-\pi}^{0} \right) = \frac{1}{2\pi} \left(0 - (-\pi) - \left(\frac{\sin 0}{2} - \frac{\sin 2\pi}{2} \right) \right)$$

$$\Rightarrow b_1 = \frac{1}{2}$$

Choice (3) is the answer.

Notes

In this problem, the relations below have been used:

$$1 - \cos 2x = 2 \sin^2 x$$

$$\int x^n dx = \frac{1}{n+1} x^{n+1}$$

$$\int \cos ax \, dx = \frac{1}{a} \sin ax$$

$$\sin 0 = 0$$

$$\sin 2\pi = 0$$

10.6. Based on the information given in the problem, we have:

$$f(x) = \begin{cases} \dfrac{-1}{2}\pi & -\pi \le x < 0 \\ \dfrac{1}{2}\pi & 0 \le x < \pi \end{cases}$$

As we know, b_3 is the coefficient of sin3x. Moreover, the coefficient of sine terms in the Fourier series of a periodic function can be calculated as follows:

$$b_n = \frac{2}{T}\int_{-\frac{T}{2}}^{\frac{T}{2}} f(x)\sin nx \; dx$$

Therefore:

$$b_3 = \frac{2}{2\pi}\int_{-\pi}^{\pi} f(x)\sin 3x \; dx$$

$$\Rightarrow b_3 = \frac{1}{\pi}\int_{-\pi}^{0} -\frac{1}{2}\pi \sin 3x \; dx + \frac{1}{\pi}\int_{0}^{\pi} \frac{1}{2}\pi \sin 3x \; dx$$

$$\Rightarrow b_3 = -\frac{1}{2}\int_{-\pi}^{0} \sin 3x \; dx + \frac{1}{2}\int_{0}^{\pi} \sin 3x \; dx$$

$$\Rightarrow b_3 = -\frac{1}{2}\left[-\frac{\cos 3x}{3}\right]_{-\pi}^{0} + \frac{1}{2}\left[-\frac{\cos 3x}{3}\right]_{0}^{\pi}$$

$$\Rightarrow b_3 = \frac{1}{6}(\cos 0 - \cos 3\pi) - \frac{1}{6}(\cos 3\pi - \cos 0)$$

$$\Rightarrow b_3 = \frac{1}{3}(\cos 0 - \cos 3\pi) = \frac{1}{3}(1 - (-1))$$

$$\Rightarrow b_3 = \frac{2}{3}$$

Choice (3) is the answer.

Notes

In this problem, the relations below have been used:

$$\int \sin ax \; dx = -\frac{1}{a}\cos ax$$

$$\cos 0 = 1$$

$$\cos 3\pi = -1$$

10.2   Half-Domain Fourier Sine and Cosine Series of Aperiodic Functions

10.7. Based on the information given in the problem, we have:

$$f(x) = \begin{cases} 1 & 0 \le x < \pi \\ 0 & \pi \le x < 2\pi \end{cases}$$

The half-domain Fourier cosine series of an aperiodic function can be calculated as follows:

$$a_0 = \frac{2}{T} \int_0^{\frac{T}{2}} f(x) dx$$

$$a_n = \frac{4}{T} \int_0^{\frac{T}{2}} f(x) \cos\left(\frac{2n\pi}{T}x\right) dx$$

$$f(x) = a_0 + \sum_{n=1}^{\infty} a_n \cos\left(\frac{2n\pi}{T}x\right)$$

Therefore:

$$a_0 = \frac{2}{4\pi} \left(\int_0^{\pi} dx + \int_{\pi}^{2\pi} 0 \times dx \right) = \frac{1}{2\pi} [x]_0^{\pi} = \frac{1}{2}$$

$$a_n = \frac{4}{4\pi} \left(\int_0^{\pi} \cos\left(\frac{n}{2}x\right) dx + \int_{\pi}^{2\pi} 0 \times \cos\left(\frac{n}{2}x\right) dx \right) = \frac{2}{n\pi} \left[\sin\left(\frac{n}{2}x\right) \right]_0^{\pi} = \frac{2}{n\pi} \sin\left(\frac{n\pi}{2}\right)$$

$$\Rightarrow f(x) = \frac{1}{2} + \sum_{n=1}^{\infty} \frac{2}{n\pi} \sin\left(\frac{n\pi}{2}\right) \cos\left(\frac{nx}{2}\right)$$

Choice (1) is the answer.

Notes

In this problem, the relations below have been used:

$$\int x^n dx = \frac{1}{n+1} x^{n+1}$$

$$\int \cos ax \, dx = \frac{1}{a} \sin ax$$

$$\sin 0 = 0$$

10.8. Based on the information given in the problem, we have:

$$f(x) = \begin{cases} 1 & 0 \leq x < \pi \\ 0 & \pi \leq x < 2\pi \end{cases}$$

The half-domain Fourier sine series of an aperiodic function can be calculated as follows:

$$b_n = \frac{4}{T} \int_0^{\frac{T}{2}} f(x) \sin\left(\frac{2n\pi}{T}x\right) dx$$

$$f(x) = \sum_{n=1}^{\infty} b_n \sin\left(\frac{2n\pi}{T}x\right)$$

Therefore:

$$b_n = \frac{4}{4\pi}\left(\int_0^{\pi} \sin\left(\frac{n}{2}x\right) dx + \int_{\pi}^{2\pi} 0 \times \sin\left(\frac{n}{2}x\right) dx\right) = -\frac{2}{n\pi}\left[\cos\left(\frac{n}{2}x\right)\right]_0^{\pi} = -\frac{2}{n\pi}\left[\cos\left(\frac{n\pi}{2}\right) - 1\right]$$

$$\Rightarrow f(x) = \sum_{n=1}^{\infty}\left(-\frac{2}{n\pi}\left[\cos\left(\frac{n\pi}{2}\right) - 1\right]\right) \sin\left(\frac{nx}{2}\right)$$

Choice (1) is the answer.

Notes

In this problem, the relations below have been used:

$$\int \sin ax \, dx = -\frac{1}{a}\cos ax$$

$$\cos 0 = 1$$

10.9. Based on the information given in the problem, we have:

$$f(x) = |\sin x|, \quad -\pi \leq x < \pi$$

$$f(x) = \frac{2}{\pi} + \frac{4}{\pi}\sum_{n=1}^{\infty} \frac{\cos 2nx}{1 - 4n^2}$$

$$g(x) = \cos x, 0 \leq x < \pi$$

As we know, for $0 \leq x < \pi$, $f(x) = \sin x$, which is the integral of $g(x) = \cos x$. In other words:

$$g(x) = \frac{d}{dx}f(x)$$

Therefore:

$$g(x) = \frac{d}{dx}\left(\frac{2}{\pi} + \frac{4}{\pi}\sum_{n=1}^{\infty}\frac{\cos 2nx}{1 - 4n^2}\right)$$

$$\Rightarrow g(x) = -\frac{8}{\pi}\sum_{n=1}^{\infty}\frac{n\sin 2nx}{(1 - 4n^2)}$$

Choice (2) is the answer.

Notes

In this problem, the relation below has been used:

$$\frac{d}{dx}\cos ax = -a\sin ax$$

10.10. Based on the information given in the problem, we have:

$$f(x) = \begin{cases} -1 & -\pi \leq x < 0 \\ 1 & 0 \leq x < \pi \end{cases},$$

$$f(x) = \frac{4}{\pi}\left[\frac{\sin x}{1} + \frac{\sin 3x}{3} + \frac{\sin 5x}{5} + \cdots\right] = \frac{4}{\pi}\sum_{k=0}^{\infty}\frac{\sin(2k+1)x}{2k+1}$$

$$g(x) = |x|, \quad -\pi \leq x < \pi$$

As can be noticed, $g(x)$ is the integral of $f(x)$. In other words:

$$g(x) = \int_0^x f(x)\,dx$$

Therefore:

$$g(x) = \int_0^x \frac{4}{\pi}\sum_{k=0}^{\infty}\frac{\sin(2k+1)x}{2k+1}\,dx$$

$$\Rightarrow g(x) = \frac{4}{\pi}\sum_{k=0}^{\infty}\int_0^x \frac{\sin(2k+1)x}{2k+1}\,dx$$

$$\Rightarrow g(x) = \frac{4}{\pi}\sum_{k=0}^{\infty}\left[-\frac{\cos(2k+1)x}{(2k+1)^2}\right]_0^x$$

$$\Rightarrow g(x) = \frac{4}{\pi}\sum_{k=0}^{\infty}\left[\frac{1 - \cos(2k+1)x}{(2k+1)^2}\right]$$

Choice (3) is the answer.

10.3   Complex Fourier Series of Periodic Functions

10.11. Based on the information given in the problem, we have:

$$f(x) = x^2, \quad 0 \le x < 2\pi$$

The complex Fourier series of a periodic function can be calculated as follows:

$$f(x) = \sum_{k=-\infty}^{\infty} C_n e^{i\frac{2\pi n}{T}x}$$

where

$$C_0 = \frac{1}{T} \int_{-\frac{T}{2}}^{\frac{T}{2}} f(x)\, dx$$

$$C_n = \frac{1}{T} \int_{-\frac{T}{2}}^{\frac{T}{2}} f(x) e^{-i\frac{2\pi n}{T}x} dx$$

Therefore:

$$C_0 = \frac{1}{2\pi} \int_0^{2\pi} x^2 dx = \frac{1}{2\pi} \left[\frac{x^3}{3} \right]_0^{2\pi} = \frac{4\pi^2}{3}$$

$$C_n = \frac{1}{2\pi} \int_0^{2\pi} x^2 e^{-inx} dx = \frac{1}{2\pi} \left[-\frac{x^2 e^{-inx}}{in} - \frac{2xe^{-inx}}{-n^2} - \frac{2e^{-inx}}{-in^3} \right]_0^{2\pi}$$

$$\Rightarrow C_n = \frac{1}{2\pi} \left[\left(-\frac{4\pi^2 e^{-in2\pi}}{in} - \frac{4\pi e^{-in2\pi}}{-n^2} - \frac{2e^{-in2\pi}}{-in^3} \right) - \left(-0 - 0 - \frac{2}{-in^3} \right) \right]$$

$$\Rightarrow C_n = \frac{1}{2\pi} \left[-\frac{4\pi^2}{in} + \frac{4\pi}{n^2} + \frac{2}{in^3} - \frac{2}{in^3} \right]$$

$$\Rightarrow C_n = \frac{1}{2\pi} \left[\frac{4i\pi^2}{n} + \frac{4\pi}{n^2} \right] = \frac{2 + i2\pi n}{n^2}$$

$$\Rightarrow f(x) = \frac{4\pi^2}{3} + \sum_{k=-\infty}^{\infty} \frac{2 + i2\pi n}{n^2} e^{inx}$$

Choice (4) is the answer.

In this problem, the relations below have been used:

$$\int x^n dx = \frac{1}{n+1}x^{n+1}$$

$$e^{-in2\pi} = \cos(n2\pi) - i\sin(n2\pi) = 1$$

$$i^2 = -1$$

$$\int u(x)v'(x)dx = u(x)v(x) - \int v(x)u'(x)dx$$

10.4 Fourier Integral of Aperiodic Functions

10.12. Based on the information given in the problem, we have:

$$f(x) = \begin{cases} \sin x & 0 \le x < \pi \\ 0 & \text{Otherwise} \end{cases}$$

The Fourier integral of an aperiodic function can be calculated as follows:

$$f(x) = \frac{1}{\pi}\int_0^\infty [A(\omega)\cos\omega x + B(\omega)\sin\omega x]d\omega$$

where

$$A(\omega) = \int_{-\infty}^\infty f(x)\cos\omega x dx$$

$$B(\omega) = \int_{-\infty}^\infty f(x)\sin\omega x dx$$

Therefore:

$$A(\omega) = \int_0^\pi \sin x \cos\omega x dx = \frac{1}{2}\int_0^\pi [\sin(1+\omega)x + \sin(1-\omega)x]dx$$

$$\Rightarrow A(\omega) = \frac{1}{2}\left[-\frac{\cos(1+\omega)x}{1+\omega} - \frac{\cos(1-\omega)x}{1-\omega}\right]_0^\pi$$

$$\Rightarrow A(\omega) = \frac{1}{2}\left[\left(-\frac{\cos(1+\omega)\pi}{1+\omega} - \frac{\cos(1-\omega)\pi}{1-\omega}\right) - \left(-\frac{\cos 0}{1+\omega} - \frac{\cos 0}{1-\omega}\right)\right]$$

$$\Rightarrow A(\omega) = \frac{1}{2}\left[\frac{\cos\omega\pi}{1+\omega} + \frac{\cos\omega\pi}{1-\omega} + \frac{1}{1+\omega} + \frac{1}{1-\omega}\right]$$

$$\Rightarrow A(\omega) = \frac{1 + \cos \omega \pi}{1 - \omega^2}$$

$$B(\omega) = \int_0^\pi \sin x \sin \omega x \, dx = \frac{1}{2} \int_0^\pi [\cos(1 - \omega)x - \cos(1 + \omega)x] dx$$

$$\Rightarrow B(\omega) = \frac{1}{2} \left[\frac{\sin(1 - \omega)x}{1 - \omega} - \frac{\sin(1 + \omega)x}{1 + \omega} \right]_0^\pi$$

$$\Rightarrow B(\omega) = \frac{1}{2} \left[\left(\frac{\sin(1 - \omega)\pi}{1 - \omega} - \frac{\sin(1 + \omega)\pi}{1 + \omega} \right) - \left(\frac{\sin 0}{1 - \omega} - \frac{\sin 0}{1 + \omega} \right) \right]$$

$$\Rightarrow B(\omega) = \frac{1}{2} \left[\left(\frac{\sin \omega \pi}{1 - \omega} - \frac{\sin \omega \pi}{1 + \omega} \right) - (0 - 0) \right]$$

$$\Rightarrow B(\omega) = -\frac{\sin \omega \pi}{1 - \omega^2}$$

$$\Rightarrow f(x) = \frac{1}{\pi} \int_0^\infty \left[\frac{1 + \cos \omega \pi}{1 - \omega^2} \cos \omega x - \frac{\sin \omega \pi}{1 - \omega^2} \sin \omega x \right] d\omega$$

Choice (1) is the answer.

Notes

In this problem, the relations below have been used:

$$\sin a \cos b = \frac{1}{2} (\sin(a + b) + \sin(a - b))$$

$$\sin a \sin b = \frac{1}{2} (\cos(a - b) - \cos(a + b))$$

$$\int \sin ax \, dx = -\frac{1}{a} \cos ax$$

$$\int \cos ax \, dx = \frac{1}{a} \sin ax$$

$$\cos(\pi + \alpha) = -\cos \alpha$$

$$\cos(\pi - \alpha) = -\cos \alpha$$

$$\cos 0 = 1$$

$$\sin(\pi + \alpha) = -\sin \alpha$$

$$\sin(\pi - \alpha) = \sin \alpha$$

$$\sin 0 = 0$$

10.13. Based on the information given in the problem, we need to calculate the value of the following integral by using the Fourier integral of the function $f(x) = e^{-x}$, $x > 0$.

$$I = \int_0^\infty \frac{\cos \omega x + \omega \sin \omega x}{1 + \omega^2} d\omega$$

As we know, the Fourier integral of an aperiodic function can be calculated as follows:

$$A(\omega) = \int_{-\infty}^\infty f(x) \cos \omega x \, dx$$

$$B(\omega) = \int_{-\infty}^\infty f(x) \sin \omega x \, dx$$

$$f(x) = \frac{1}{\pi} \int_0^\infty [A(\omega) \cos \omega x + B(\omega) \sin \omega x] \, d\omega$$

Therefore, for the given function, we have:

$$A(\omega) = \int_0^\infty e^{-x} \cos \omega x \, dx$$

Herein, the Laplace transform definition can be used as follows:

$$A(\omega) = L\{\cos \omega x\}\Big|_{s=1} = \frac{s}{s^2 + \omega^2}\Big|_{s=1} \Rightarrow A(\omega) = \frac{1}{1 + \omega^2}$$

$$B(\omega) = \int_0^\infty e^{-x} \sin \omega x \, dx$$

Again, the Laplace transform definition can be used as follows:

$$B(\omega) = L\{\sin \omega x\}\Big|_{s=1} = \frac{\omega}{s^2 + \omega^2}\Big|_{s=1} \Rightarrow B(\omega) = \frac{\omega}{1 + \omega^2}$$

$$f(x) = \frac{1}{\pi} \int_0^\infty \left[\frac{1}{1 + \omega^2} \cos \omega x + \frac{\omega}{1 + \omega^2} \sin \omega x\right] d\omega = \frac{1}{\pi} \int_0^\infty \frac{\cos \omega x + \omega \sin \omega x}{1 + \omega^2} d\omega$$

Hence:

$$e^{-x} = \frac{1}{\pi} \int_0^\infty \frac{\cos \omega x + \omega \sin \omega x}{1 + \omega^2} d\omega$$

$$\Rightarrow \int_0^\infty \frac{\cos \omega x + \omega \sin \omega x}{1 + \omega^2} d\omega = \pi e^{-x}$$

$$\Rightarrow I = \pi e^{-x}$$

Choice (3) is the answer.

Notes

In this problem, the relations below have been used:

$$L\{f(x)\} = \int_0^\infty f(x)e^{-sx}dx$$

$$L\{\cos \omega x\} = \frac{s}{s^2 + \omega^2}$$

$$L\{\sin \omega x\} = \frac{\omega}{s^2 + \omega^2}$$

10.5   Complex Fourier Integral of Aperiodic Functions

10.14. Based on the information given in the problem, we have:

$$f(x) = \begin{cases} \sinh 2x & 0 \leq x < 2 \\ 0 & x \geq 2 \end{cases}$$

The complex Fourier integral of an aperiodic function can be calculated as follows:

$$f(x) = \int_{-\infty}^{+\infty} C(\omega)e^{i\omega x}d\omega$$

where

$$C(\omega) = \frac{1}{2\pi}\int_{-\infty}^{+\infty} f(x)e^{-i\omega x}dx$$

Thus:

$$C(\omega) = \frac{1}{2\pi}\int_0^2 e^{-i\omega x}\sinh 2x\ dx = \frac{1}{2\pi}\int_0^2 e^{-i\omega x}\frac{e^{2x} - e^{-2x}}{2}dx$$

$$\Rightarrow C(\omega) = \frac{1}{4\pi}\int_0^2 \left[e^{(2-i\omega)x} - e^{-(2+i\omega)x}\right]dx$$

$$\Rightarrow C(\omega) = \frac{1}{4\pi}\left[\frac{e^{(2-i\omega)x}}{2 - i\omega} + \frac{e^{-(2+i\omega)x}}{2 + i\omega}\right]_0^2$$

$$\Rightarrow C(\omega) = \frac{1}{4\pi}\left[e^{-i2\omega}\left(\frac{e^4}{2 - i\omega} + \frac{e^{-4}}{2 + i\omega}\right) - \left(\frac{1}{2 - i\omega} + \frac{1}{2 + i\omega}\right)\right]$$

$$\Rightarrow C(\omega) = \frac{1}{4\pi}\left[e^{-i2\omega}\left(\frac{2e^4 + i\omega e^4 + 2e^{-4} - i\omega e^{-4}}{4 + \omega^2}\right) - \frac{4}{4 + \omega^2}\right]$$

$$\Rightarrow C(\omega) = \frac{1}{4\pi}\left[e^{-i2\omega}\left(\frac{2\left(e^4 + e^{-4}\right) + i\omega\left(e^4 - e^{-4}\right)}{4 + \omega^2}\right) - \frac{4}{4 + \omega^2}\right]$$

$$\Rightarrow C(\omega) = \frac{1}{4\pi}\left[e^{-i2\omega}\left(\frac{4\cosh 4 + i2\omega\sinh 4}{4 + \omega^2}\right) - \frac{4}{4 + \omega^2}\right]$$

$$\Rightarrow C(\omega) = \frac{1}{4\pi}\left[\frac{e^{-i2\omega}(4\cosh 4 + i2\omega\sinh 4) - 4}{4 + \omega^2}\right]$$

$$\Rightarrow f(x) = \frac{1}{4\pi}\int_{-\infty}^{+\infty}\left[\frac{e^{-i2\omega}(4\cosh 4 + i2\omega\sinh 4) - 4}{4 + \omega^2}\right]e^{i\omega x}d\omega$$

Choice (4) is the answer.

Notes

In this problem, the relations below have been used:

$$\sinh a = \frac{e^a - e^{-a}}{2}$$

$$\cosh a = \frac{e^a + e^{-a}}{2}$$

$$\int e^{ax}dx = \frac{1}{a}e^{ax}$$

$$e^{a+b} = e^a e^b$$

10.6   Fourier Transform of Aperiodic Functions

10.15. Based on the information given in the problem, we have:

$$f(x) = \begin{cases} 1 & |x| \le a \\ 0 & |x| > a \end{cases}$$

The Fourier transform and inverse Fourier transform of an aperiodic function can be calculated, respectively, as follows:

$$F(\omega) = \int_{-\infty}^{+\infty} f(x)e^{-i\omega x}\,dx$$

$$f(x) = \frac{1}{2\pi}\int_{-\infty}^{+\infty} f(x)e^{i\omega x}\,d\omega$$

Therefore:

$$F(\omega) = \int_{-a}^{a} e^{-i\omega x}dx = \left[-\frac{1}{i\omega}e^{-i\omega x}\right]_{-a}^{a}$$

$$\Rightarrow F(\omega) = -\frac{1}{i\omega}\left(e^{-i\omega a} - e^{i\omega a}\right) = -\frac{1}{i\omega}\left(-2i\sin\omega a\right)$$

$$\Rightarrow F(\omega) = \frac{2\sin\omega a}{\omega}$$

Choice (1) is the answer.

Notes

In this problem, the relations below have been used:

$$\int e^{ax}dx = \frac{1}{a}e^{ax}$$

$$\sin a = \frac{e^{ia} - e^{-ia}}{2i}$$

10.16. Based on the information given in the problem, we have:

$$y'' + 4y = \begin{cases} 0 & x \le 0 \\ e^{-2x} & x > 0 \end{cases}$$

The problem can be solved by calculating the Fourier transform of each side of the differential equation and using the Laplace transform definition as follows:

$$(i\omega)^2 Y(\omega) + 4Y(\omega) = \int_0^\infty e^{-i\omega x} e^{-2x} dx$$

$$\Rightarrow (4 - \omega^2) Y(\omega) = L\{e^{-2x}\}\Big|_{s = i\omega}$$

$$\Rightarrow (4 - \omega^2) Y(\omega) = \frac{1}{s+2}\Big|_{s = i\omega}$$

$$\Rightarrow (4 - \omega^2) Y(\omega) = \frac{1}{i\omega + 2}$$

$$\Rightarrow Y(\omega) = \frac{1}{(i\omega + 2)(4 - \omega^2)}$$

Choice (2) is the answer.

Notes

In this problem, the relations below have been used:

$$F(\omega) = \int_{-\infty}^{+\infty} f(x)e^{-i\omega x}dx$$

$$F\left\{f^{(n)}(x)\right\} = (i\omega)^n F(\omega)$$

$$i^2 = -1$$

$$L\{f(x)\} = \int_0^{\infty} f(x)e^{-sx}dx$$

$$L\{e^{-ax}\} = \frac{1}{s+a}$$

10.17. Based on the information given in the problem, we have:

$$y' - 4y = \begin{cases} e^{-4x} & x \geq 0 \\ 0 & x < 0 \end{cases}$$

The problem can be solved by calculating the Fourier transform of each side of the differential equation and using the Laplace transform definition as follows:

$$i\omega Y(\omega) - 4Y(\omega) = \int_0^{\infty} e^{-i\omega x}e^{-4x}dx$$

$$\Rightarrow i\omega Y(\omega) - 4Y(\omega) = L\{e^{-4x}\}\Big|_{s = i\omega}$$

$$\Rightarrow i\omega Y(\omega) - 4Y(\omega) = \frac{1}{s+4}\Big|_{s = i\omega}$$

$$\Rightarrow i\omega Y(\omega) - 4Y(\omega) = \frac{1}{i\omega + 4}$$

$$\Rightarrow Y(\omega) = \left(\frac{1}{i\omega + 4}\right)\left(\frac{1}{i\omega - 4}\right) = \frac{1}{(i\omega)^2 - 4^2}$$

$$\Rightarrow Y(\omega) = \frac{-1}{\omega^2 + 16}$$

Choice (1) is the answer.

In this problem, the relations below have been used:

$$F(\omega) = \int_{-\infty}^{+\infty} f(x)e^{-i\omega x}\,dx$$

$$F\left\{f^{(n)}(x)\right\} = (i\omega)^n F(\omega)$$

$$i^2 = -1$$

$$L\{f(x)\} = \int_0^\infty f(x)e^{-sx}\,dx$$

$$L\{e^{-ax}\} = \frac{1}{s+a}$$

10.7   Half-Domain Fourier Sine and Cosine Transforms of Aperiodic Functions

10.18. Based on the information given in the problem, the Fourier cosine series of the aperiodic function $f(x) = x, 0 \leq x < \pi$ is given as follows:

$$f(x) = \frac{\pi}{2} - \frac{4}{\pi}\left(\cos x + \frac{1}{3^2}\cos 3x + \frac{1}{5^2}\cos 5x + \cdots\right) \tag{1}$$

Moreover, we have:

$$g(x) = x(\pi - x)\frac{\pi}{8}, 0 \leq x < \pi \tag{2}$$

From (1), we can write:

$$x = \frac{\pi}{2} - \frac{4}{\pi}\sum_{n=1}^{\infty}\frac{\cos(2n-1)x}{(2n-1)^2} \tag{3}$$

$$\Rightarrow x - \frac{\pi}{2} = -\frac{4}{\pi}\sum_{n=1}^{\infty}\frac{\cos(2n-1)x}{(2n-1)^2} \tag{4}$$

By integrating each side of the previous equation, we have:

$$\frac{x^2}{2} - \frac{\pi}{2}x + c = -\frac{4}{\pi}\sum_{n=1}^{\infty}\frac{\sin(2n-1)x}{(2n-1)^3}$$

The value of the constant c is determined as zero by calculating the limit of each side of the previous equation. Hence:

$$\frac{x}{2}(x - \pi) = -\frac{4}{\pi} \sum_{n=1}^{\infty} \frac{\sin(2n-1)x}{(2n-1)^3}$$

$$\Rightarrow x(\pi - x)\frac{\pi}{8} = \sum_{n=1}^{\infty} \frac{\sin(2n-1)x}{(2n-1)^3}$$

Choice (4) is the answer.

Notes

In this problem, the relations below have been used:

$$\int x^n dx = \frac{1}{n+1} x^{n+1}$$

$$\int \cos ax \, dx = \frac{1}{a} \sin ax$$

References

1. Rahmani-Andebili, M. (2024). Precalculus (2nd Ed.) – Practice Problems, Methods, and Solutions, Springer Nature.
2. Rahmani-Andebili, M. (2023). Calculus III – Practice Problems, Methods, and Solutions, Springer Nature.
3. Rahmani-Andebili, M. (2023). Calculus II – Practice Problems, Methods, and Solutions, Springer Nature.
4. Rahmani-Andebili, M. (2023). Calculus I (2nd Ed.) – Practice Problems, Methods, and Solutions, Springer Nature.
5. Rahmani-Andebili, M. (2022). Differential Equations – Practice Problems, Methods, and Solutions, Springer Nature.
6. Rahmani-Andebili, M. (2021). Calculus – Practice Problems, Methods, and Solutions, Springer Nature.
7. Rahmani-Andebili, M. (2021). Precalculus – Practice Problems, Methods, and Solutions, Springer Nature.

Abstract

In this chapter, the basic and advanced problems of partial differential equations are presented. The subjects include determining the type of partial differential equations, updating partial differential equations by new variables, solving partial differential equations by using the techniques used in solving ordinary differential equations, solving partial differential equations by using their characteristics equations, solving partial differential equations by using variables separation method, solving partial differential equations by Laplace transform, and solving partial differential equations in a steady-state condition. Herein, different types of problems and exercises are presented that are categorized as follows:

○ **Problems with detailed solution**: They have been designed to teach students the subjects in detail. Moreover, they have been categorized into different levels based on their difficulty levels (easy, normal, and hard) and calculation amounts (small, normal, and large).

○ **Partially solved exercises**: They have been designed to encourage students to practice more problems while guiding them through the problem-solving procedure and hinting the required formulas.

○ **Exercises with final answer**: They have been designed to encourage students to practice by themselves while hinting them by the final answer as well as to help instructors to give tests or quizzes.

11.1 Type of Partial Differential Equations

11.1 Which one of the following choices is true about the type of the partial differential equation below [1–7]?

$$x\frac{\partial^2 u(x, y)}{\partial x^2} + y\frac{\partial^2 u(x, y)}{\partial y^2} + 3y^2\frac{\partial u(x, y)}{\partial x} = 0$$

Difficulty level   ○ Easy   ● Normal   ○ Hard
Calculation amount   ● Small   ○ Normal ○ Large

1) In the region $xy < 0$, it is hyperbolic, and in the region $xy > 0$, it is parabolic.
2) In the region $xy < 0$, it is parabolic, and in the region $xy > 0$, it is elliptic.
3) In the region $xy < 0$, it is hyperbolic, and in the region $xy > 0$, it is elliptic.
4) In the region $xy < 0$, it is elliptic, and in the region $xy > 0$, it is hyperbolic.

Partially Solved Exercise

Which one of the following choices is true about the type of the partial differential equation below?

$$yu_{xx} + u_{yy} = 0$$

1) For $y > 0$, it is elliptic.
2) For $y < 0$, it is parabolic.
3) For $y > 0$, it is hyperbolic.
4) For $y = 0$, it is hyperbolic.

Solution

The discriminant of a homogeneous second-order partial differential equation can help determine its type. Consider the general form of the homogeneous second-order partial differential equation presented in the following:

$$Au_{xx} + Bu_{xy} + Cu_{yy} + Du_x + Eu_y + Fu = 0$$

The discriminant of the homogeneous second-order partial differential equation is defined as follows:

$$\Delta = B^2 - 4AC$$

If $\Delta < 0$, the equation is classified as elliptic.
If $\Delta = 0$, the equation is classified as parabolic.
If $\Delta > 0$, the equation is classified as hyperbolic.

Therefore, for the given equation, we can write:

$$\Delta = (\quad)^2 - 4(\quad)(\quad) = -4y$$

Hence, for $y > 0$, the value of the discriminant is negative, and the equation is elliptic. Choice (1) is the answer.

Exercise

What is the type of the partial differential equation $xu_{xx} + yu_{yy} + 3y^2u_x = 0$ for $=y$?

Final Answer
Elliptic.

11.2. Which one of the following choices is true regarding the type of following partial differential equation?

$$\frac{\partial^2 u(x, y)}{\partial x^2} + 6\frac{\partial^2 u(x, y)}{\partial x \partial y} + 5\frac{\partial^2 u(x, y)}{\partial y^2} = 0$$

Difficulty level   ○ Easy   ● Normal   ○ Hard
Calculation amount   ● Small   ○ Normal   ○ Large

1) It is hyperbolic.
2) It is elliptic.
3) It is parabolic.
4) Its type is unknown.

Partially Solved Exercise

Consider the following partial differential equation:

$$(x-1)\frac{\partial^2 u(x,y)}{\partial x^2} + 2(y+1)\frac{\partial^2 u(x,y)}{\partial x \partial y} - (x+1)\frac{\partial^2 u(x,y)}{\partial y^2} = 0$$

What is the type of the partial differential equation in the region $y > 0$?
1) Elliptic
2) Parabolic
3) Hyperbolic
4) Spherical

Solution

The discriminant of a homogeneous second-order partial differential equation can help determine its type. Consider the general form of the homogeneous second-order partial differential equation below:

$$A(x,y)\frac{\partial^2 u(x,y)}{\partial x^2} + B(x,y)\frac{\partial^2 u(x,y)}{\partial x \partial y} + C(x,y)\frac{\partial^2 u(x,y)}{\partial y^2} + D(x,y)\frac{\partial u(x,y)}{\partial x}$$
$$+ E(x,y)\frac{\partial u(x,y)}{\partial y} + F(x,y)u(x,y) = 0$$

The discriminant of the homogeneous second-order partial differential equation is defined as follows:

$$\Delta = B^2 - 4AC$$

If $\Delta < 0$, the equation is classified as elliptic.
If $\Delta = 0$, the equation is classified as parabolic.
If $\Delta > 0$, the equation is classified as hyperbolic.

Therefore, for the given equation, we can write:

$$\Delta = (\qquad)^2 - 4(\qquad)(\qquad)$$

$$\Rightarrow \Delta =$$

$$\Rightarrow \Delta = 4x^2 + 4y^2 + 8y$$

$$\xrightarrow{y>0} \Delta > 0$$

Thus, the equation is hyperbolic. Choice (3) is the answer.

11.3. What is the type of following partial differential equation on the curve $y = \frac{x^2}{1+x}$, $x \neq -1$?

$$(x+1)u_{xx} + 2xu_{xy} + yu_{yy} = u$$

Difficulty level   ○ Easy   ○ Normal   ● Hard
Calculation amount   ● Small   ○ Normal   ○ Large
1) Elliptic
2) Parabolic
3) Hyperbolic
4) Spherical

11.2 Updating Partial Differential Equations by New Variables

11.4. Consider the partial differential equation below:

$$x\frac{\partial u(x,y)}{\partial x} + y\frac{\partial u(x,y)}{\partial x} = nu(x,y)$$

Update the partial differential equation by using the new variables $p = \frac{1}{2}(\ln x + \ln y)$ and $q = \frac{1}{2}(\ln x - \ln y)$.
Difficulty level   ○ Easy   ● Normal   ○ Hard
Calculation amount   ○ Small   ○ Normal   ● Large
1) $\dfrac{\partial u(p,q)}{\partial p} = nu(p,q)$

2) $\dfrac{\partial u(p,q)}{\partial q} = nu(p,q)$

3) $\dfrac{\partial u(p,q)}{\partial p} + \dfrac{\partial u(p,q)}{\partial q} = nu(p,q)$

4) $\dfrac{\partial u(p,q)}{\partial p} - \dfrac{\partial u(p,q)}{\partial q} = nu(p,q)$

11.5. Consider the following partial differential equation:

$$\frac{\partial u(t,s)}{\partial s} = 0$$

Convert the partial differential equation to a new one by using the new variables $t = x + 2y$ and $s = x$.
Difficulty level   ○ Easy   ● Normal   ○ Hard
Calculation amount   ○ Small   ● Normal   ○ Large
1) $2\dfrac{\partial u(x,y)}{\partial x} - \dfrac{\partial u(x,y)}{\partial y} = 0$

2) $\dfrac{\partial u(x,y)}{\partial x} + \dfrac{\partial u(x,y)}{\partial y} = 0$

3) $2\dfrac{\partial u(x,y)}{\partial x} + \dfrac{\partial u(x,y)}{\partial y} = 0$

4) $\dfrac{\partial u(x,y)}{\partial x} + 2\dfrac{\partial u(x,y)}{\partial y} = 0$

11.3 Solving Partial Differential Equations by Using the Techniques Used in Solving Ordinary Differential Equations

11.6. Solve the partial differential equation below:

$$t\frac{\partial^2 u(x,t)}{\partial x \partial t} + 2\frac{\partial u(x,t)}{\partial x} = x^2$$

Difficulty level   ○ Easy   ○ Normal   ● Hard
Calculation amount   ○ Small   ● Normal   ○ Large

1) $u(x,t) = tx^2 + h(t) + g(x)$
2) $u(x,t) = t^3x^3 + h(t) + g(x)$
3) $u(x,t) = \frac{1}{6}t^2x^2 + h(t) + \frac{g(x)}{t^2}$
4) $u(x,t) = \frac{x^3}{6} + h(t) + \frac{g(x)}{t^2}$

11.4 Solving Partial Differential Equations by Using Their Characteristics Equations

11.7. Which one of the following choices is the general solution of the partial differential equation below?

$$2\frac{\partial^2 u(x,y)}{\partial x^2} + 4\frac{\partial^2 u(x,y)}{\partial x \partial y} + \frac{\partial^2 u(x,y)}{\partial y^2} = 0$$

Difficulty level   ○ Easy   ○ Normal   ● Hard
Calculation amount   ● Small   ○ Normal   ○ Large

1) $f\left(\frac{y+x}{\sqrt{2}} - y\right) + g\left(\frac{y-x}{\sqrt{2}} + y\right)$
2) $f\left(y - 1 + \frac{x}{\sqrt{2}}\right) + g\left(y - 1 + \frac{x}{\sqrt{2}}\right)$
3) $f\left(y + \left(\frac{1}{\sqrt{2}} - 1\right)x\right) + g\left(y - \left(\frac{1}{\sqrt{2}} + 1\right)x\right)$
4) $f\left(y - 1 + \frac{x}{\sqrt{2}}\right) + g\left(y - x - \frac{x}{\sqrt{2}}\right)$

Partially Solved Exercise

Find the general solution of the partial differential equation below:

$$\frac{\partial^2 u(x,y)}{\partial x^2} - \frac{\partial^2 u(x,y)}{\partial x \partial y} - 2\frac{\partial^2 u(x,y)}{\partial y^2} = 0$$

Solution

To find the general solution of the partial differential equation in the form $Au_{xx} + Bu_{xy} + Cu_{yy} = 0$, we need to determine the roots of the characteristic equation concerned with the partial differential equation as follows:

$$A\lambda^2 + B\lambda + C = 0 \Rightarrow \lambda = \lambda_1, \lambda_2$$

If the roots are real and distinct, the general solution of the partial differential equation is as follows:

$$u(x,y) = f(y + \lambda_1 x) + g(y + \lambda_2 x)$$

If the roots are real but the same, the general solution of the partial differential equation is as follows:

$$u(x,y) = f(y + \lambda x) + xg(y + \lambda x)$$

If the roots are imaginary and complex conjugate of each other, the general solution of the partial differential equation is in one of the following forms:

$$u(x,y) = f(y + \lambda x) + g(y + \lambda^* x)$$

$$u(x,y) = f(x + \lambda y) + g(x + \lambda^* y)$$

Therefore, for this problem, we have:

$$\lambda^2 + (\quad)\lambda + (\quad) = 0$$

$$\Rightarrow \lambda = \frac{(\quad) \pm \sqrt{(\quad) + (\quad)}}{(\quad)} \Rightarrow \lambda_{1,2} = (\quad), (\quad)$$

As can be seen, the roots are (   ). Thus:

$$u(x,y) = g(y - x) + h(y + 2x)$$

Exercise

Find the general solution of the partial differential equation below:

$$u_{xx}(x,y) + 2u_{xy}(x,y) + u_{yy}(x,y) = 0$$

Final Answer

$u(x,y) = f(y + x) + xg(y + x)$

Exercise

Find the general solution of the partial differential equation below:

$$u_{xx}(x,y) + u_{yy}(x,y) = 0$$

Final Answer

$u(x, y) = f(y + ix) + g(y - ix)$

11.5 Solving Partial Differential Equations by Using Variables Separation Method

11.8. Solve the partial differential equation below:

$$u_x(x, y) + u_y(x, y) = 2(x + y)u(x, y)$$

Difficulty level   ○ Easy   ○ Normal   ● Hard
Calculation amount   ○ Small   ● Normal   ○ Large

1) $u(x, y) = ce^{x^2 + y^2 + k(x-y)}$
2) $u(x, y) = ce^{x^2 - y^2 + k(x-y)}$
3) $u(x, y) = ce^{x^2 + y^2 + k(x+y)}$
4) $u(x, y) = ce^{x^2 - y^2 + k(x+y)}$

11.6 Solving Partial Differential Equations by Laplace Transform

11.9. Apply Laplace transform to present the following partial differential equation in Laplace domain:

$$\frac{\partial u(x, t)}{\partial x} + \frac{\partial u(x, t)}{\partial t} + u(x, t) = xt, \, u(x, 0) = 0$$

Difficulty level   ○ Easy   ● Normal   ○ Hard
Calculation amount   ● Small   ○ Normal   ○ Large

1) $\dfrac{\partial U(x, s)}{\partial t} + sU(x, s) = \dfrac{x}{s^2}$

2) $\dfrac{\partial U(x, s)}{\partial t} - (s + 1)U(x, s) = \dfrac{x}{t^2}$

3) $\dfrac{\partial U(x, s)}{\partial x} - (s + 1)U(x, s) = \dfrac{x}{s^2}$

4) $\dfrac{\partial U(x, s)}{\partial x} + (s + 1)U(x, s) = \dfrac{x}{s^2}$

11.7 Solving Partial Differential Equations in Steady-State Condition

11.10. A one-dimensional nonhomogeneous heat conduction problem is as follows:

$$\frac{\partial^2 u(x, t)}{\partial x^2} + a = \frac{\partial u(x, t)}{\partial t}, \quad 0 \le x < L, \quad t > 0$$

$$u(0, t) = T_0, \quad t > 0$$

$$\frac{\partial u(L,t)}{\partial x}=0, \quad t>0$$

Calculate the steady-state value of the problem solution (steady-state temperature of conductor).

Difficulty level   ○ Easy   ● Normal   ○ Hard
Calculation amount   ○ Small   ● Normal   ○ Large

1) $ax(L-x)+T_0$
2) $\frac{ax}{2}(2L-x)+T_0$
3) $\frac{ax}{2}(L-x)^2+T_0$
4) $\frac{ax}{2}(L-x)+T_0$

Partially Solved Exercise

A one-dimensional nonhomogeneous heat conduction problem is as follows:

$$\frac{\partial u(x,t)}{\partial t}=\frac{\partial^2 u(x,t)}{\partial x^2}+1, \quad 0\le x<L, \quad t>0 \tag{1}$$

$$u(0,t)=u(L,t)=0 \tag{2}$$

Calculate the steady-state value of the solution of the problem (steady-state temperature of conductor).

Solution

The steady-state value of the solution of a problem $(u_{ss}(x)=\lim_{t\to\infty} u(x,t))$ can be calculated by assuming that:

$$\frac{\partial u(x,t)}{\partial t}\bigg|_{t\to\infty}=\frac{\partial u_{ss}(x)}{\partial t}=0 \tag{3}$$

Solving (1) and (3) in the steady-state condition:

$$0=\frac{\partial^2 u_{ss}(x)}{\partial x^2}+1 \Rightarrow \frac{\partial^2 u_{ss}(x)}{\partial x^2}= \tag{4}$$

$$\frac{\int dx}{\Rightarrow}\ \frac{\partial u_{ss}(x)}{\partial x}= \tag{5}$$

$$\frac{\int dx}{\Rightarrow}\ u_{ss}(x)= \tag{6}$$

Solving (2) and (6):

$$\begin{cases} u_{ss}(0)=0 \\ u_{ss}(L)=0 \end{cases} \Rightarrow \begin{cases} 0= \\ 0= \end{cases}$$

$$\Rightarrow k_1 = , k_2 =$$

$$\Rightarrow u_{ss}(x) =$$

$$\Rightarrow u_{ss}(x) = \frac{x(L - x)}{2}$$

Notes

In this problem, the relation below has been used:

$$\int x^n dx = \frac{x^{n+1}}{n + 1} + c$$

Exercise

A one-dimensional homogeneous heat conduction problem is as follows:

$$u_t = c^2 u_{xx}, \quad 0 \leq x < L$$

$$u(0, t) = a$$

$$u(L, t) = b$$

Calculate the steady-state value of the solution of the problem (steady-state temperature of conductor).

Final Answer

$(b - a)\frac{x}{L} + a$

Exercise

A one-dimensional homogeneous heat conduction problem is as follows:

$$\frac{\partial u(x, t)}{\partial t} = c^2 \frac{\partial^2 u(x, t)}{\partial x^2}, \quad 0 \leq x < L$$

$$u(0, t) = T_0$$

$$u(L, t) = 2T_0$$

Calculate the steady-state value of the solution of the problem (steady-state temperature of conductor).

Final Answer

$T_0\left(1 + \frac{x}{L}\right)$

11.11. A one-dimensional homogeneous heat conduction problem is as follows:

$$\frac{\partial u(x,t)}{\partial t} = \frac{\partial^2 u(x,t)}{\partial x^2}, \quad 0 \leq x < 1, \quad t > 0$$

$$u(1,t) = 2$$

$$u(0,t) = 1$$

Calculate the steady-state value of the problem solution (steady-state temperature of conductor) at $x = \frac{2}{3}$.

Difficulty level   ○ Easy   ● Normal   ○ Hard

Calculation amount   ○ Small   ● Normal   ○ Large

1) $\frac{3}{2}$

2) $\frac{4}{3}$

3) $\frac{5}{3}$

4) $\frac{13}{9}$

Exercise

A one-dimensional nonhomogeneous heat conduction problem is as follows:

$$\frac{\partial^2 u(x,t)}{\partial x^2} - \frac{1}{c^2}\frac{\partial u(x,t)}{\partial t} = 1, \quad 0 \leq x < L, \quad t > 0$$

$$u(0,t) = u(L,t) = 0$$

Calculate the steady-state value of the problem solution (steady-state temperature of conductor) at $x = \frac{1}{3}L$.

Final Answer

$-\frac{L^2}{9}$

Exercise

A one-dimensional nonhomogeneous heat conduction problem is as follows:

$$\frac{\partial^2 u(x,t)}{\partial x^2} - \frac{\partial u(x,t)}{\partial t} = 1, \quad 0 \leq x < 1, \quad t > 0$$

$$u(0,t) = u(1,t) = 0$$

Calculate the steady-state value of the problem solution (steady-state temperature of conductor) at $x = 0.5$.

Final Answer

$-\frac{1}{8}$

References

1. Rahmani-Andebili, M. (2024). Precalculus (2nd Ed.) – Practice Problems, Methods, and Solutions, Springer Nature.
2. Rahmani-Andebili, M. (2023). Calculus III – Practice Problems, Methods, and Solutions, Springer Nature.
3. Rahmani-Andebili, M. (2023). Calculus II – Practice Problems, Methods, and Solutions, Springer Nature.
4. Rahmani-Andebili, M. (2023). Calculus I (2nd Ed.) – Practice Problems, Methods, and Solutions, Springer Nature.
5. Rahmani-Andebili, M. (2022). Differential Equations – Practice Problems, Methods, and Solutions, Springer Nature.
6. Rahmani-Andebili, M. (2021). Calculus – Practice Problems, Methods, and Solutions, Springer Nature.
7. Rahmani-Andebili, M. (2021). Precalculus – Practice Problems, Methods, and Solutions, Springer Nature.

Abstract

In this chapter, the problems of the 11th chapter are fully solved, in detail, step-by-step, and with different methods.

12.1 Type of Partial Differential Equations

12.1. Based on the information given in the problem, we have [1–7]:

$$x\frac{\partial^2 u(x,y)}{\partial x^2} + y\frac{\partial^2 u(x,y)}{\partial y^2} + 3y^2\frac{\partial u(x,y)}{\partial x} = 0$$

The discriminant of a homogeneous second-order partial differential equation can help determine its type. Consider the general form of the homogeneous second-order partial differential equation presented below:

$$A(x,y)\frac{\partial^2 u(x,y)}{\partial x^2} + B(x,y)\frac{\partial^2 u(x,y)}{\partial x \partial y} + C(x,y)\frac{\partial^2 u(x,y)}{\partial y^2} + D(x,y)\frac{\partial u(x,y)}{\partial x}$$
$$+ E(x,y)\frac{\partial u(x,y)}{\partial y} + F(x,y)u(x,y) = 0$$

The discriminant of the homogeneous second-order partial differential equation is defined as follows:

$$\Delta = B^2 - 4AC$$

If $\Delta < 0$, the equation is classified as elliptic.
If $\Delta = 0$, the equation is classified as parabolic.
If $\Delta > 0$, the equation is classified as hyperbolic.

Therefore, for the given equation, we can write:

$$\Delta = 0^2 - 4(x)(y) = -4xy$$

Thus, the equation in the region $xy < 0$ is hyperbolic (because $\Delta > 0$), and in the region $xy > 0$, it is elliptic (because $\Delta < 0$). Choice (3) is the answer.

12.2. Based on the information given in the problem, we have:

$$\frac{\partial^2 u(x,y)}{\partial x^2} + 6\frac{\partial^2 u(x,y)}{\partial x \partial y} + 5\frac{\partial^2 u(x,y)}{\partial y^2} = 0$$

The discriminant of a homogeneous second-order partial differential equation can help determine its type. Consider the general form of the homogeneous second-order partial differential equation below:

$$A(x,y)\frac{\partial^2 u(x,y)}{\partial x^2} + B(x,y)\frac{\partial^2 u(x,y)}{\partial x \partial y} + C(x,y)\frac{\partial^2 u(x,y)}{\partial y^2} + D(x,y)\frac{\partial u(x,y)}{\partial x}$$
$$+ E(x,y)\frac{\partial u(x,y)}{\partial y} + F(x,y)u(x,y) = 0$$

The discriminant of the homogeneous second-order partial differential equation is defined as follows:

$$\Delta = B^2 - 4AC$$

If $\Delta < 0$, the equation is classified as elliptic.
If $\Delta = 0$, the equation is classified as parabolic.
If $\Delta > 0$, the equation is classified as hyperbolic.

Therefore, for the given equation, we can write:

$$\Delta = (6)^2 - 4 \times 1 \times 5 = 16 > 0$$

Hence, the equation is hyperbolic. Choice (1) is the answer.

12.3. Based on the information given in the problem, we have:

$$y = \frac{x^2}{1+x}, \quad x \neq -1 \tag{1}$$

$$(x+1)u_{xx} + 2xu_{xy} + yu_{yy} = u \tag{2}$$

The discriminant of a homogeneous second-order partial differential equation can help determine its type. Consider the general form of the homogeneous second-order partial differential equation below:

$$Au_{xx} + Bu_{xy} + Cu_{yy} + Du_x + Eu_y + Fu = 0$$

The discriminant of the homogeneous second-order partial differential equation is defined as follows:

$$\Delta = B^2 - 4AC$$

If $\Delta < 0$, the equation is classified as elliptic.
If $\Delta = 0$, the equation is classified as parabolic.
If $\Delta > 0$, the equation is classified as hyperbolic.

Therefore, for the given equation, we can write:

$$\Delta = (2x)^2 - 4(x+1)y \tag{3}$$

$$\Rightarrow \Delta = 4\left(x^2 - xy - y\right) = 4\left(x^2 - y(x+1)\right) \tag{4}$$

Solving (1) and (4):

$$\Delta = 4\left(x^2 - \frac{x^2}{1+x}(x+1)\right)$$

$$\Rightarrow \Delta = 0$$

Thus, the equation is parabolic. Choice (2) is the answer.

12.2 Updating Partial Differential Equations by New Variables

12.4. Based on the information given in the problem, we have:

$$x\frac{\partial u(x,y)}{\partial x} + y\frac{\partial u(x,y)}{\partial x} = nu(x,y) \tag{1}$$

$$\begin{cases} p(x,y) = \dfrac{1}{2}\left(\ln x + \ln y\right) \\ q(x,y) = \dfrac{1}{2}\left(\ln x - \ln y\right) \end{cases} \tag{2}$$

The partial differential equation can be updated based on the new variables by using the chain rule as follows:

$$\begin{cases} \dfrac{\partial u(x,y)}{\partial x} = \dfrac{\partial u(p,q)}{\partial p}\dfrac{\partial p(x,y)}{\partial x} + \dfrac{\partial u(p,q)}{\partial q}\dfrac{\partial q(x,y)}{\partial x} \\ \dfrac{\partial u(x,y)}{\partial y} = \dfrac{\partial u(p,q)}{\partial p}\dfrac{\partial p(x,y)}{\partial y} + \dfrac{\partial u(p,q)}{\partial q}\dfrac{\partial q(x,y)}{\partial y} \end{cases} \tag{3}$$

$$\begin{cases} \dfrac{\partial p(x,y)}{\partial x} = \dfrac{1}{2x} \\ \dfrac{\partial q(x,y)}{\partial x} = \dfrac{1}{2x} \\ \dfrac{\partial p(x,y)}{\partial y} = \dfrac{1}{2y} \\ \dfrac{\partial q(x,y)}{\partial y} = -\dfrac{1}{2y} \end{cases} \tag{4}$$

Solving (3) and (4):

$$\begin{cases} \dfrac{\partial u(x,y)}{\partial x} = \dfrac{\partial u(p,q)}{\partial p}\dfrac{1}{2x} + \dfrac{\partial u(p,q)}{\partial q}\dfrac{1}{2x} \\ \dfrac{\partial u(x,y)}{\partial y} = \dfrac{\partial u(p,q)}{\partial p}\dfrac{1}{2y} + \dfrac{\partial u(p,q)}{\partial q}\left(-\dfrac{1}{2y}\right) \end{cases} \tag{5}$$

Solving (1) and (5):

$$x\left(\frac{\partial u(p,q)}{\partial p}\frac{1}{2x}+\frac{\partial u(p,q)}{\partial q}\frac{1}{2x}\right)+y\left(\frac{\partial u(p,q)}{\partial p}\frac{1}{2y}+\frac{\partial u(p,q)}{\partial q}\left(-\frac{1}{2y}\right)\right)=nu(p,q)$$

$$\Rightarrow \frac{1}{2}\frac{\partial u(p,q)}{\partial p}+\frac{1}{2}\frac{\partial u(p,q)}{\partial q}+\frac{1}{2}\frac{\partial u(p,q)}{\partial p}-\frac{1}{2}\frac{\partial u(p,q)}{\partial q}=nu(p,q)$$

$$\Rightarrow \frac{\partial u(p,q)}{\partial p}=nu(p,q)$$

Choice (1) is the answer.

Notes

In this problem, the relations below have been used:

$$\frac{\partial u}{\partial x}=\frac{\partial u}{\partial p}\frac{\partial p}{\partial x}+\frac{\partial u}{\partial q}\frac{\partial q}{\partial x}$$

$$\frac{\partial u}{\partial y}=\frac{\partial u}{\partial p}\frac{\partial p}{\partial y}+\frac{\partial u}{\partial q}\frac{\partial q}{\partial y}$$

$$\frac{d}{dx}\ln x=\frac{1}{x}$$

12.5. Based on the information given in the problem, we have:

$$\frac{\partial u(t,s)}{\partial s}=0 \tag{1}$$

$$\begin{cases}t=x+2y\\s=x\end{cases}\Rightarrow\begin{cases}y=\frac{1}{2}(t-s)\\x=s\end{cases} \tag{2}$$

The partial differential equation can be updated based on the new variables by using the chain rule as follows:

$$\frac{\partial u(t,s)}{\partial s}=\frac{\partial u(x,y)}{\partial x}\frac{\partial x(t,s)}{\partial s}+\frac{\partial u(x,y)}{\partial y}\frac{\partial y(t,s)}{\partial s} \tag{3}$$

$$\begin{cases}\dfrac{\partial x(t,s)}{\partial s}=1\\[4mm]\dfrac{\partial y(t,s)}{\partial s}=-\dfrac{1}{2}\end{cases} \tag{4}$$

Solving (3) and (4):

$$\frac{\partial u(t,s)}{\partial s}=\frac{\partial u(x,y)}{\partial x}-\frac{1}{2}\frac{\partial u(x,y)}{\partial y} \tag{5}$$

Solving (1) and (5):

$$\frac{\partial u(x,y)}{\partial x} - \frac{1}{2}\frac{\partial u(x,y)}{\partial y} = 0$$

$$\Rightarrow 2\frac{\partial u(x,y)}{\partial x} - \frac{\partial u(x,y)}{\partial y} = 0$$

Choice (1) is the answer.

Notes

In this problem, the relation below has been used:

$$\frac{\partial u}{\partial s} = \frac{\partial u}{\partial x}\frac{\partial x}{\partial s} + \frac{\partial u}{\partial y}\frac{\partial y}{\partial s}$$

12.3 Solving Partial Differential Equations by Using the Techniques Used in Solving Ordinary Differential Equations

12.6. Based on the information given in the problem, we have:

$$t\frac{\partial^2 u(x,t)}{\partial x \partial t} + 2\frac{\partial u(x,t)}{\partial x} = x^2$$

By integrating the equation with respect to x, we have:

$$t\frac{\partial u(x,t)}{\partial t} + 2u(x,t) = \frac{x^3}{3} + f(t)$$

Herein, $f(t)$ is the constant value of integration with respect to x.

$$\overset{\times t}{\Rightarrow} t^2\frac{\partial u(x,t)}{\partial t} + 2tu(x,t) = \frac{x^3}{3}t + tf(t)$$

$$\Rightarrow \frac{\partial}{\partial t}\left(t^2 u(x,t)\right) = \frac{x^3}{3}t + tf(t)$$

By integrating the equation with respect to t, we have:

$$t^2 u(x,t) = \frac{x^3 t^2}{6} + \int tf(t)dt + g(x)$$

Herein, $g(x)$ is the constant value of integration with respect to t.

$$\xrightarrow{\times \frac{1}{t^2}} u(x,t) = \frac{x^3}{6} + \frac{1}{t^2}\int tf(t)dt + \frac{1}{t^2}g(x)$$

By assuming $h(t) = \frac{1}{t^2} \int tf(t)dt$, we have:

$$u(x,t) = \frac{x^3}{6} + h(t) + \frac{g(x)}{t^2}$$

Choice (4) is the answer.

Notes

In this problem, the relations below have been used:

$$\int \left(v(x)\frac{d}{dx}u(x) + u(x)\frac{d}{dx}v(x) \right)dx = u(x)v(x) + c$$

$$\int x^n dx = \frac{x^{n+1}}{n+1} + c$$

12.4   Solving Partial Differential Equations by Using Their Characteristics Equations

12.7. Based on the information given in the problem, we have:

$$2\frac{\partial^2 u(x,y)}{\partial x^2} + 4\frac{\partial^2 u(x,y)}{\partial x \partial y} + \frac{\partial^2 u(x,y)}{\partial y^2} = 0$$

To find the general solution of the partial differential equation in the form $Au_{xx} + Bu_{xy} + Cu_{yy} = 0$, we need to determine the roots of the characteristic equation concerned with the partial differential equation as follows:

$$A\lambda^2 + B\lambda + C = 0 \Rightarrow \lambda = \lambda_1, \lambda_2$$

If the roots are real and distinct, the general solution of the partial differential equation is as follows:

$$u(x,y) = f(y + \lambda_1 x) + g(y + \lambda_2 x)$$

If the roots are real but the same, the general solution of the partial differential equation is as follows:

$$u(x,y) = f(y + \lambda x) + xg(y + \lambda x)$$

If the roots are imaginary and complex conjugate of each other, the general solution of the partial differential equation is in one of the following forms:

$$u(x,y) = f(x + \lambda y) + g(x + \lambda^* y)$$

$$u(x,y) = f(y + \lambda x) + g(y + \lambda^* x)$$

Therefore, for this problem, we have:

$$2\lambda^2 + 4\lambda + 1 = 0$$

$$\Rightarrow \lambda = \frac{-4 \pm \sqrt{16-8}}{4} = \frac{-2 \pm \sqrt{2}}{2} \Rightarrow \lambda_{1,2} = \left(\frac{1}{\sqrt{2}} - 1\right), \quad -\left(\frac{1}{\sqrt{2}} + 1\right)$$

As can be seen, the roots are real and distinct. Thus:

$$u(x, y) = f\left(y + \left(\frac{1}{\sqrt{2}} - 1\right)x\right) + g\left(y - \left(\frac{1}{\sqrt{2}} + 1\right)x\right)$$

Choice (3) is the answer.

12.5   Solving Partial Differential Equations by Using Variables Separation Method

12.8. Based on the information given in the problem, we have:

$$u_x(x, y) + u_y(x, y) = 2(x + y)u(x, y) \tag{1}$$

The problem can be solved by variables separation. Herein, it is assumed that $u(x, y)$ can be written as follows:

$$u(x, y) = F(x)G(y) \tag{2}$$

Solving (1) and (2):

$$F'(x)G(y) + F(x)G'(y) = (2x + 2y)F(x)G(y) \tag{3}$$

$$\Rightarrow \frac{F'(x)}{F(x)} + \frac{G'(y)}{G(y)} = 2x + 2y \tag{4}$$

$$\Rightarrow \frac{F'(x)}{F(x)} - 2x = -\frac{G'(y)}{G(y)} + 2y \tag{5}$$

As can be seen in (5), each side of the equation is based on a specific variable. Thus, each side of the equation must be equal to a constant value. Therefore:

$$\begin{cases} \dfrac{F'(x)}{F(x)} - 2x = k \Rightarrow \dfrac{F'(x)}{F(x)} = 2x + k \\[2mm] -\dfrac{G'(y)}{G(y)} + 2y = k \Rightarrow \dfrac{G'(y)}{G(y)} = 2y - k \end{cases} \tag{6}$$

$$\begin{cases} \xRightarrow{\int dx} \ln F(x) - \ln c_1 = x^2 + kx \\[2mm] \xRightarrow{\int dy} \ln F(y) - \ln c_2 = y^2 - ky \end{cases} \tag{7}$$

$$\Rightarrow \begin{cases} \ln \dfrac{F(x)}{c_1} = x^2 + kx \Rightarrow F(x) = c_1 e^{x^2 + kx} \\[2mm] \ln \dfrac{G(y)}{c_2} = y^2 - ky \Rightarrow G(y) = c_2 e^{y^2 - ky} \end{cases} \tag{8}$$

Solving (2) and (8):

$$u(x, y) = c_1 c_2 e^{x^2 + y^2 + k(x-y)}$$

$$\Rightarrow u(x, y) = c e^{x^2 + y^2 + k(x-y)}$$

Choice (1) is the answer.

Notes

In this problem, the relations below have been used:

$$\int \frac{du}{u} = u + c$$

$$\int x^n dx = \frac{x^{n+1}}{n+1} + c$$

$$\ln a - \ln b = \ln \frac{a}{b}$$

$$\ln a = b \Rightarrow a = e^b$$

$$e^a e^b = e^{a+b}$$

12.6   Solving Partial Differential Equations by Using Laplace Transform

12.9. Based on the information given in the problem, we have:

$$\frac{\partial u(x, t)}{\partial x} + \frac{\partial u(x, t)}{\partial t} + u(x, t) = xt \tag{1}$$

$$u(x, 0) = 0 \tag{2}$$

By transferring to Laplace domain, we have:

$$\frac{\partial U(x, s)}{\partial x} + (sU(x, s) - u(x, 0)) + U(x, s) = \frac{x}{s^2} \tag{3}$$

Herein, we considered that:

$$L\{u(x, t)\} = U(x, s) \tag{4}$$

$$L\left\{\frac{\partial u(x, t)}{\partial t}\right\} = sU(x, s) - u(x, 0) \tag{5}$$

$$L\{t\} = \frac{1}{s^2} \tag{6}$$

Solving (2) and (3):

$$\frac{\partial U(x,s)}{\partial x} + (s+1)U(x,s) = \frac{x}{s^2}$$

Choice (4) is the answer.

Notes

In this problem, the relation below has been used:

$$L\{t^n\} = \frac{n!}{s^{n+1}}$$

12.7   Solving Partial Differential Equations in Steady-State Condition

12.10. Based on the information given in the problem, we have:

$$\frac{\partial^2 u(x,t)}{\partial x^2} + a = \frac{\partial u(x,t)}{\partial t}, \quad 0 \le x < L, \quad t > 0 \tag{1}$$

$$u(0,t) = T_0, \quad t > 0 \tag{2}$$

$$\frac{\partial u(L,t)}{\partial x} = 0, \quad t > 0 \tag{3}$$

The steady-state value of the solution of a problem $(u_{ss}(x) = \lim_{t \to \infty} u(x,t))$ can be calculated by assuming that:

$$\frac{\partial u(x,t)}{\partial t}\bigg|_{t \to \infty} = \frac{\partial u_{ss}(x)}{\partial t} = 0 \tag{4}$$

Solving (1) and (4) in the steady-state condition:

$$\frac{\partial^2 u_{ss}(x)}{\partial x^2} + a = 0 \tag{5}$$

$$\Rightarrow \frac{\partial^2 u_{ss}(x)}{\partial x^2} = -a \tag{6}$$

$$\xrightarrow{\int dx} \frac{\partial u_{ss}(x)}{\partial x} = -ax + k_1 \tag{7}$$

$$\xrightarrow{\int dx} u_{ss}(x) = -\frac{a}{2}x^2 + k_1 x + k_2 \tag{8}$$

Solving (2) and (8):

$$T_0 = -0 + 0 + k_2 \Rightarrow k_2 = T_0 \tag{9}$$

Solving (3), (8), and (9):

$$0 = \frac{\partial}{\partial x}\left(-\frac{a}{2}x^2 + k_1 x + T_0\right)\bigg|_{x=L} \tag{10}$$

$$\Rightarrow 0 = (-ax + k_1)|_{x=L} \tag{11}$$

$$\Rightarrow 0 = -aL + k_1 \Rightarrow k_1 = aL \tag{12}$$

Solving (8), (9), and (12):

$$u_{ss}(x) = -\frac{a}{2}x^2 + aLx + T_0$$

$$\Rightarrow u_{ss}(x) = \frac{ax}{2}(2L - x) + T_0$$

Choice (2) is the answer.

Notes

In this problem, the relation below has been used:

$$\int x^n dx = \frac{x^{n+1}}{n+1} + c$$

12.11. Based on the information given in the problem, we need to calculate the steady-state value of the solution of the problem (steady-state temperature of conductor) at $x = \frac{2}{3}$. Moreover, we have:

$$\frac{\partial u(x,t)}{\partial t} = \frac{\partial^2 u(x,t)}{\partial x^2}, \quad 0 \le x < 1, \quad t > 0 \tag{1}$$

$$u(1,t) = 2 \tag{2}$$

$$u(0,t) = 1 \tag{3}$$

The steady-state value of the solution of a problem $(u_{ss}(x) = \lim_{t \to \infty} u(x,t))$ can be calculated by assuming that:

$$\frac{\partial u(x,t)}{\partial t}\bigg|_{t \to \infty} = \frac{\partial u_{ss}(x)}{\partial t} = 0 \tag{4}$$

Solving (1) and (4) in the steady-state condition:

$$\frac{\partial^2 u_{ss}(x)}{\partial x^2} = 0 \tag{5}$$

$$\xrightarrow{\int dx} \frac{\partial u_{ss}(x)}{\partial x} = k_1 \tag{6}$$

$$\xrightarrow{\int dx} u_{ss}(x) = k_1 x + k_2 \tag{7}$$

Solving (2), (3), and (7):

$$\begin{cases} 2 = k_1 + k_2 \\ 1 = k_2 \end{cases} \Rightarrow k_1 = 1, \quad k_2 = 1 \tag{8}$$

$$\Rightarrow u_{ss}(x) = x + 1 \tag{9}$$

$$\Rightarrow u_{ss}\left(\frac{2}{3}\right) = \frac{2}{3} + 1 \tag{10}$$

$$\Rightarrow u_{ss}\left(\frac{2}{3}\right) = \frac{5}{3}$$

Choice (3) is the answer.

Notes

In this problem, the relation below has been used:

$$\int x^n dx = \frac{x^{n+1}}{n+1} + c$$

References

1. Rahmani-Andebili, M. (2024). Precalculus (2nd Ed.) – Practice Problems, Methods, and Solutions, Springer Nature.
2. Rahmani-Andebili, M. (2023). Calculus III – Practice Problems, Methods, and Solutions, Springer Nature.
3. Rahmani-Andebili, M. (2023). Calculus II – Practice Problems, Methods, and Solutions, Springer Nature.
4. Rahmani-Andebili, M. (2023). Calculus I (2nd Ed.) – Practice Problems, Methods, and Solutions, Springer Nature.
5. Rahmani-Andebili, M. (2022). Differential Equations – Practice Problems, Methods, and Solutions, Springer Nature.
6. Rahmani-Andebili, M. (2021). Calculus – Practice Problems, Methods, and Solutions, Springer Nature.
7. Rahmani-Andebili, M. (2021). Precalculus – Practice Problems, Methods, and Solutions, Springer Nature.

Index

A
Aperiodic function, 174–178, 187–189, 191–199

C
Cartesian coordinates, 70
Cauchy-Riemann equations, 20, 21, 47
Cauchy's integral theorem, 129, 145, 146
Characteristics equation, 205–207, 218–219
Circle, 14, 16–19, 37, 38, 40, 42, 43, 45, 55, 58–64, 70–76, 78–82, 127–129, 132–135, 138, 139, 144–146, 149, 150, 152, 154, 155, 157, 160, 162–165
Clockwise, 68, 130–134, 138, 141, 147–149, 151, 153, 154, 156, 158, 159, 161, 163, 164, 167, 168
Complex conjugate, 206, 218
Complex equation, 11–20, 32–46
Complex Fourier integral, 171–178, 194–195
Complex Fourier series, 171–178, 190–191
Complex functions, 1–23, 25–51, 89–124, 128, 129, 132, 133, 135, 139, 145–155, 157–162, 164–169
Complex integration, 125–141, 143–169
Complex quantities, 1–10, 26, 27
Complex series, 89–124
Complex transformations, 53–65, 67–87
Contour, 127–132, 134, 136, 138, 140, 141, 145–154, 156–159, 161, 163–169
Convergence radius, 93, 94, 108–110
Convergence region, 94, 111
Counterclockwise, 128–132, 134–139, 141, 145–149, 151–154, 156, 157, 159, 161, 163, 164, 166–168
Curve, 125, 204

D
Direction of contour, 129–131, 133, 135–137, 139, 141, 145, 146, 149, 151, 154, 156, 159, 161, 164, 168
Discriminant, 202, 203, 213, 214
Distinct, 206, 218, 219

E
Ellipse, 17–19, 61, 64
Elliptic, 201–204, 213, 214
Essential singularity, 89–93
Euler's formula, 13, 34, 35
Exponential, 13, 91, 106, 121
Exponential complex transformation, 61–63, 79–82

F
Finite number of singular points, 130–141, 147–169
Fourier integral, 171–178, 191–194
Fourier series, 171–178, 181–199
Fourier transform, 171–178, 195–198

G
General solution, 205, 206, 218
Geometric sequence, 28

H
Half-domain Fourier sine and cosine series, 171–178, 187–189
Half-domain Fourier sine and cosine transforms, 178, 198–199
Harmonic conjugate function, 20–23, 47–51
Harmonic function, 22, 23, 48–50
Heat conduction problem, 207–210
Holomorphic function, 20, 21, 47, 112, 113, 115, 116, 128–130, 145–147
Homogeneous, 202, 203, 209, 210, 213, 214
Hyperbola, 17, 19, 44, 64, 85, 86
Hyperbolic, 201–204, 213, 214
Hyperbolic sine and cosine complex transformation, 64, 84–85

I
Image, 53–64, 70–72
Imaginary, 15, 47, 59, 206, 218
Integral, 125–141, 145–149, 151, 153, 154, 156–159, 161, 163, 164, 166–168, 177, 188, 189, 193

L
Laplace transform, 193, 196, 197, 207, 220–221
Laurent series expansion, 89–124, 133, 135, 138, 139, 148–150, 152, 153, 155–158, 160–164, 166, 168, 169
Left half plane, 54
Limit, 1–23, 25–51, 92, 104–107, 199
Line, 18, 19, 37, 56, 58–61, 64, 105
Linear complex transformation, 53–55
Linear fractional complex transformation, 65, 86–87
Lower half plane, 14

N
Natural logarithm, 8, 9, 63
Natural logarithm complex transformation, 63–64, 82–83